Strategic Management of Built Facilities

Strategic Management of Built Facilities

Craig Langston
and Rima Lauge-Kristensen

OXFORD AUCKLAND BOSTON JOHANNESBURG MELBOURNE NEW DELHI

Butterworth-Heinemann
Linacre House, Jordan Hill, Oxford OX2 8DP
225 Wildwood Avenue, Woburn, MA 01801-2041

A division of Reed Educational and Professional Publishing Ltd

First published 2002

British Library Cataloguing in Publication Data
A catalogue record for this book is available from the British Library

Library of Congress Cataloguing in Publication Data
A catalogue record for this book is available from the Library of Congress

ISBN 0 7506 5440 6

For information on all Butterworth-Heinemann publications visit our website at www.bh.com

Typeset by Avocet Typeset, Brill, Aylesbury, Bucks
Printed and bound in Great Britain

Contents

List of contributors

Editors

Craig Langston – Associate Professor in Construction Economics at the University of Technology, Sydney (UTS). Before commencing at UTS, he worked for a professional quantity surveying office in Sydney. His PhD thesis was concerned with discounting and life-cost studies. He developed two cost-planning software packages (PROPHET and LIFECOST) that are sold internationally, and is author of several books covering aspects of construction economics and facility management.

Rima Lauge-Kristensen – Architect and research scholar. She holds a PhD in physics and a bachelor of arts degree in architecture, and has worked in a number of roles relating to the construction industry. She has particular expertise in the physics of building systems and emerging technologies that aim to minimize energy demand.

Specialist contributions

Rick Best – Senior Lecturer in Construction Economics at UTS. He has degrees in architecture and quantity surveying and has research interests in information technology, energy in buildings and low energy design strategies. Rick is undertaking research in district energy systems and the international performance of construction projects, and is the co-editor of several books on value in building.

Gerard de Valence – Senior Lecturer in Construction Economics at UTS. He has an honours degree in economics from the University of Sydney. He has worked in industry as an analyst and economist. His principal areas of research interest include the measurement of project performance, economic factors related to the construction industry, and the impact of emerging technologies.

Grace Ding – Lecturer in Construction Economics at UTS. She has a diploma from Hong Kong Polytechnic, a bachelor's degree from the University of Ulster and a master's degree by thesis from the University of Salford. She has practised as a quantity surveyor in Hong Kong, England and Australia. Grace is currently undertaking PhD research in the area of environmental performance measurement.

Peter Smith – Senior Lecturer in Construction Economics at UTS. Prior to his current appointment he worked for a large professional quantity surveying practice and an international construction and property development company. He is currently undertaking PhD research and has an interest in consumer investment advice and property maintenance.

John Twyford – Senior Lecturer in Construction Economics at UTS. He has a doctorate in judicial science and is editor of the *Australian Construction Law Newsletter*. He has significant industry experience in the development and administration of construction contracts and has been involved in educating students in this field for more than 20 years.

Foreword

Wes McGregor

Only the future lies ahead. The strategic imperative of all enterprises that aspire to success is to view the future as an opportunity. An opportunity to develop and then implement plans leading to sustainable comparative advantage over their competitors.

Strategic management is the means of guiding an enterprise through its actions to its chosen future. It involves determining purpose and objectives; formulating plans of action to achieve these objectives, most probably in a climate of change; implementing the actions; then evaluating the outcome against the objectives and making changes where required. Strategic management has for long been an established component of business, yet its application in the management of key resources of business – its property and related facilities services – has often been conspicuous by its absence. Why?

Ranging as it does from maintenance to security, and from cleaning to workplace planning, facilities management is more often than not perceived as a low-grade activity of no real consequence to business. Concomitant with this, facilities managers are generally perceived as having a low level of contribution to the success of the organization. Nothing could be further from the reality. The built environment for work (workspace) is rarely, if ever, neutral in its influence upon the performance of the enterprise. It either helps or it hinders. It either supports or it inhibits. It either stimulates or it demotivates.

Workspace and its facilities infrastructure are the tools at the disposal of the people in the business to enable them to achieve the enterprise's goals. The key aim of their strategic management is to add value to the organization through the effective management of the built environment. Yet the perception of business leaders is all too often that facilities and their managers are resource consumers rather than value adders. Changing this view is a major challenge faced by all facilities managers.

The performance of facilities managers impacts upon the performance of their organizations' workspace, which itself has a direct impact upon the performance of workers and hence that of the enterprise as a whole. Through their actions, facilities managers have a vital part to play in enhancing corporate performance and thereby in delivering shareholder value. Some business leaders have responded by outsourcing facilities and their management, believing that this will lead to reduced operating costs which in turn will enhance shareholder value. However, strategic management should not, indeed some would say (and I would not disagree) cannot, be outsourced. By its very nature it is the strategic vision that makes each enterprise unique.

Changing the focus of facilities managers from cost reduction and operational to tactical and strategic – i.e. forward looking and not being guided by a 'rear view mirror' approach – is at the core of the required transition. With crystal clarity, the goal to which all facilities managers should aspire is to migrate their operations from being *reactive* to *proactive* as they progress towards *anticipatory* management. In so doing, they become value adding through helping to support improvements in individual and corporate performance. There can be few if any businesses that have not recognized, and then had to come to terms with the implications, that their success is critically dependent upon the human capital of their organization. Focusing upon the recruitment and retention of staff is but the first step of recognition. The greater challenge to be faced is in liberating the potential creative power of individuals and teams through providing enabling and supportive work environments.

Strategies forged in a vacuum are doomed to fail, and hence are worthless. Strategies with no clear plan for their implementation are of no greater value. Therefore, just as the development of related tactical plans is a vital component of the realization of the strategy, so too is the development of appropriate performance measures. For many this will necessitate establishing new evaluation criteria and metrics which correlate the performance of workspace, and its facilities, to the performance of workers and the enterprise.

To do so requires data on workplace performance and the knowledge and skills to know what to do in response to its critical analysis. The use of real time data capture in which the condition and operating performance of workspace is freely available is made increasingly possible by the deployment of the tools of e-business. Open channel remote monitoring and control of facilities on a 24 by 7 basis, and 'hands off' procurement operated directly by empowered users, are only two of the many internet-based approaches which hold-out the prospect to revolutionize the use and management of facilities. In the coming years e-business and its counterpart e-facilities will radically change the nature and accessibility of facilities and services, and in so doing will yet again bring into question the role of those responsible for their management.

Strategic management is future oriented. There was no divine ordinance issued to create workspace, and I suspect none will be decreed to bring about its demise. Instead, workspace has evolved and will continue to evolve in response to corporate and human needs. The one certainty is that the need for workspace, and consequently its facilities infrastructure, will continue to change. It is, in my opinion, likely to reduce in favour of more innovative business models that employ empowerment of individuals to adopt workstyles that are much less geographically dependent, and much more instinctive, intuitive and above all easy to use. What then the role of the facilities manager? How does his or her role today help the enterprise prepare itself for a world which is less certain and less well defined?

An effective response cannot be fashioned in isolation. It requires close collaboration with all support service professionals – those responsible for human resources, information technology and finance, as well as those charged with the management of real estate and facilities. Such a model has the potential to put the facilities manager as the heart of the management of the enterprise's support infrastructure. For such an aspiration to be realized calls for facilities managers to develop new skills, acquire new tools, apply new performance measures and innovate new approaches to workspace and facilities provision. Each of these poses a discrete challenge for the individual manager. Collectively they possess the ability to elevate the facilities manager to a pivotal role in the enterprise, or alternatively to render the role obsolete and redundant. High on the list of criteria by which the future existence of facilities managers will be determined is his or her ability to apply strategic management to built facilities.

Wes McGregor is a leading workstyles and workplace consultant, a founding Director of Advanced Workplace Associates and co-author of the book *Facilities Management and the Business of Space* (Butterworth-Heinemann, 1999)

Preface

Buildings, as hosts to collective activity, have been part of the human heritage at least since the development of agricultural activity. They have expressed commercial, religious, political, aesthetic and military activity of countless societies and organizations, and equally, as Dovey (1999)[1] reminds us, have expressed and re-enforced the prevailing power structures and paradigms of those bodies. Yet it is barely twenty years since Facility Management began to emerge as a body of understanding concerning *'the practice of co-ordinating the physical workplace with the people and work of the organization'* and seeking to integrate aspects of business administration, property development and valuation, architecture and the behavioural and engineering sciences.

Facilities Management and Facility Management are frequently confused, especially in England, and countries where FM's development saw more of a British influence. As a matter of record, the earliest use of 'Facilities' does seem to pre date 'Facility' and refers to the practice of banks outsourcing the data processing operations associated with credit card processing in the 1970s. Although IT outsourcing disappeared from the Facility/Facilities discourse during the early 1980s the outsourcing tradition survived. Some would say it dominated to the point where the development of new professional knowledge in Facility Management atrophied and other terms[2] now compete for the strategic high ground. FM is in danger of being reduced to little more than maintenance management.

Craig Langston, Rima Lauge-Kristensen and their specialist contributors do much to redress the balance firmly associating Facility Management with the **strategic** management of built facilities; i.e. their management in such a way as to contribute to a particular organization's purpose and intentions. This does not mean those organizations should own those facilities. If they can be more flexible, or have core competencies which will achieve a higher return on capital, owning property may be a flawed commercial judgement. It does mean that those who manage those facilities need to consider their role, not simply or even primarily in terms of operational cost, but in terms of how they contribute to strategy. Regrettably, such outlooks are still the exception rather than the rule. Facility Managers complain in private at their exclusion from the high tables, but fail to raise their sights and thinking to strategic levels. They remain operational and think of their 'customers' as the rest of their organization rather than those who pay it to stay in business.

As economies move from land, physical labour and even capital to knowledge as the dominant factor of production, a claim can be made (Ward and Holtham, 2000)[3] that the workplace is the most under-utilized tool of contemporary knowledge management. The claim was foreseen twenty years ago by FM's intellectual founders and encapsulated in Becker's (1990)[4] workplace ecology. Ten years later a majority of built environment professionals and practitioners has failed to see the point. It is a pleasure for me to welcome a textbook that seeks to redress the balance and put the operational detail within the umbrella of strategy and the workplace ecology.

If Price, Professor of FM, Facilities Management Graduate Centre, Sheffield Hallam University, Sheffield S1 1WP, UK

---

[1] Dovey, K. *Framing Places: Mediating power in built form* (London: Routledge, 1999).

[2] For example infrastructure management, asset management and corporate real estate management.

[3] Ward, V. and Holtham, C. W. *The Role of Private and Public Spaces in Knowledge Management.* Knowledge Management: Concepts and Controversies (Conference Proceedings Only. University of Warwick, 2000).

[4] Becker, F. *The Total Workplace: Facilities Management and Elastic Organisation* (New York: Praeger Press, 1990).

Introduction

Much has been written about the design, procurement and maintenance of buildings and other infrastructure, but relatively little about the relationship between these facilities and the business functions they are intended to support. An often underestimated link, physical assets can affect the productivity of workers and the external image transmitted by the organization to its customers. Creation of facilities that have a positive influence on productivity and marketing can directly contribute to financial return and the securing of new business opportunities.

Staffing is usually the single major cost of operating a business and over any reasonable period of time could account for 80–90% of total expenditure. Therefore finding ways to reduce staff levels by streamlining processes, using technology and making workers more efficient will deliver significant recurrent savings. Facilities that are designed to minimize unnecessary travel/circulation and enable centralized supervision and resources lead to less unproductive time and duplication. Not only can people use their available time more efficiently, but often the gross area of buildings can be reduced, which impacts on a whole range of operational issues such as cleaning, energy, maintenance, and replacement. It is worthwhile devoting time to this issue during the original design phase.

However, much can be achieved after facilities are constructed to improve the way in which they function. This includes (but is not limited to) flexible and shared use of space, introduction of technology to enable people to communicate more effectively, energy-saving strategies, support services such as childcare and cafeterias, controllable internal climate and lighting control, stylish fit-out choices, and opportunities for working in remote locations that lower the need for fixed accommodation. Some of these concepts don't cost money to achieve and all of them have the potential to increase productivity and thereby enhance profits. Flexibility of work practices and good environmental citizenship can also lead to image benefits and increased market share.

Facility management is a discipline that operates at a number of levels. At its most basic, it concerns operational activities designed to ensure that facilities are maintained to a satisfactory standard, provide an appropriate working environment and are compliant with statutory requirements. It can also provide tactical input, analysing performance and looking at ways to enact tangible improvements. But most important, there is a strategic level which forms a link to core business activity so that facilities are aligned to organizational directions and managed more proactively. It is this last level that separates the routine from the dynamic, and justifies the emergence of facility management as a professional discipline in its own right.

Although there is a clear management focus necessary in all activities related to facilities, the discipline is significantly underpinned by a knowledge of economics and the ability to perform analytical studies. Known as facility economics, it is directed mainly at a strategic level to ensure that value for money is provided. Value involves improving quality, reducing cost and minimizing risk and is a combination of tangible factors (like financial return and productivity) and intangible factors (like comfort and image). An ability to review, interpret and act upon actual performance to find better ways of doing things at equal or higher quality is the key to success.

Facility management is driven by a range of new developments. These include computer-aided graphic-based databases, environmental assessment, outsourcing, benchmarking and flexible space planning. It is a worldwide phenomenon that has arisen from the significance of investment in the built environment and the need to find competitive advantage by doing things better in less time and for less money. The necessary skills to be a successful facility manager are related to communication, finance, personnel and forward planning, with a focus on customer service. Facility management is a growing profession and one with global characteristics.

The chapters in this book are categorized in a matrix arrangement to illustrate their interrelationship. The order of chapters reflects a progression from macro issues to micro issues. The vertical and horizontal integration of the chapters is shown in the diagram below.

Vertical integration is used to differentiate macro and micro issues. Chapters 1 to 12 cover strategic planning topics, while chapters 13 to 24 cover facility design and management topics. Horizontal integration is used to differentiate environmental and economic issues. Chapters 1 to 6 and 13 to 18 cover topics about organizational responsibilities, while chapters 7 to 12 and 19 to 24 cover topics about decision-making tools. Furthermore, chapters are grouped into eight parts and categorized by common themes.

Strategic planning concerns those aspects of productivity improvement that are about making effective decisions to align facilities with corporate goals and support core business tasks. Topic areas include corporate goals, functional plans, information management and risk management, and are relevant to the identification of targets and setting up systems to ensure that they are delivered. Arguably this is the essence of facility performance, and can contribute positively to workplace ecology and performance. The first half of the book concludes with recommendations on how risk management can help to future-proof an organization and provide confidence to make strategic decisions related to facility infrastructure.

Facility design and management concerns those aspects of productivity improvement that relate to underpinning processes. Topics include property maintenance, financial management, value management and building quality assessment, and make a critical contribution to the operational achievement of functional performance expectations. The second half of the book concludes by arguing that facility managers must routinely monitor and evaluate quality benchmarks if they are to pursue continuous improvement ideals.

The strength of this book is in its ability to present a balanced argument and to offer solutions that are both effective and practical. The chapters are written from an international perspective and therefore are applicable to readers from any part of the world.

This book forms a useful introductory text for construction industry professionals, facility managers, construction clients and students. The layout of the book reflects the structure of the *Facility Economics* subject in the Master of Facility Management and Master of Business Administration (Facility Management) degrees at the University of Technology, Sydney. Graduates of these courses are employed in a range of fields including engineering, architecture, property management, construction management, project management, quantity surveying and facility management, and apply their skills to projects in countries throughout the world.

The editors wish to thank all those people who contributed to the writing of this book, and to those who gave up their valuable time to provide input and/or review individual chapters.

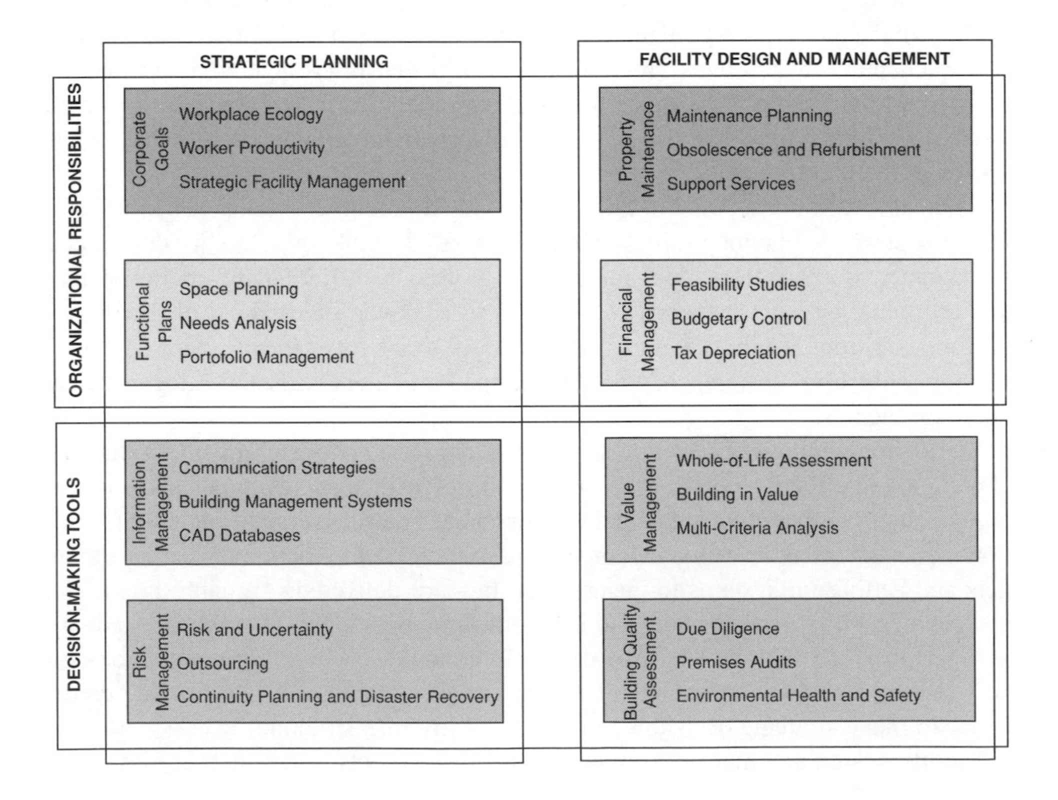

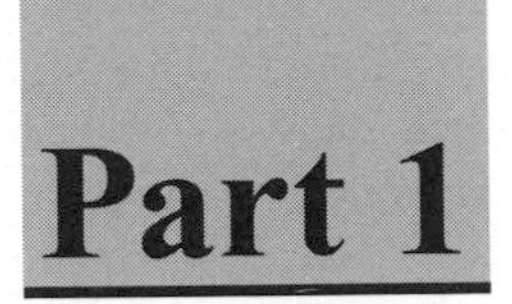

Corporate Goals

Business today is a competitive and customer-focused activity that must operate in an environment that is subject to continual change. Being successful means that the business has a clear vision and works towards identified goals while simultaneously improving quality, reducing costs and minimizing risk. Often this results in a concentration on core business tasks, with support tasks aligned to reflect overall corporate goals and strategy. More importantly, it involves empowering staff and enabling them to be productive through proper workplace design.

Facility management is about ensuring that infrastructure supports core business tasks. Infrastructure is not limited to buildings and floor space, but includes issues such as technology, communication strategy, workplace design and ergonomics, auxiliary services, security and environmental impact. In essence, facilities are the setting within which a planned activity can occur. Therefore the nature of facilities differs according to business function, industry, location, time and potential future directions. While facilities are important and in many cases represent significant capital value, they are secondary compared to human resources, despite being closely related.

The link between facility management and human resource management is worker productivity. Constructing a setting that can sustain planned activities is necessary, yet developing ways in which business processes can be more effectively performed is a higher order goal. It involves an understanding of human needs and practices from a range of stakeholder perspectives including investor, employer, employee and customer.

To be effective, all parts of the organization must be co-ordinated so that a common purpose is maintained. This is accomplished through a clearly articulated set of corporate goals and a team approach, particularly among upper management. Corporate goals usually are about core business tasks. For example, a goal is not to construct a new building but rather to introduce a new product or service to the market, which by implication requires supporting infrastructure. The role of the facility manager is to match corporate goals with the physical settings to support them, while the role of the human resource manager is to match corporate goals with the necessary skilled people for their achievement. Obviously the facility manager and the human resource manager must work closely together.

Strategic facility management represents higher order activities involved in the alignment of facilities with corporate goals. This includes a measure of anticipation and forward planning and demands access to boardroom-level discussions and/or effective communication between senior personnel. It is distinct from tactical and operational activities that have progressively reduced scope and corporate impact.

Tactical activities are essentially about monitoring and management of facility performance. For example, the conduct of an energy audit for a facility based on evidence of rising energy usage (and cost) that leads to an analysis of several new options and a recommendation for change is a tactical activity. They differ from strategic activities in that the direct link to corporate goals is absent, although the overall objective of improving quality, reducing cost and minimizing risk remains. Tactical activities are often smaller and tightly defined, and are akin to sub-projects within the greater business context.

Operational activities are more routine and are often viewed as custodial. For example, the regular schedule of filter cleaning in cooling towers, the recording of completion and the feedback of data for future planning purposes is an operational activity. They differ from tactical activities in that they do not involve analysis and judgement, but nevertheless are essential to the proper functioning of infrastructure which has obvious ramifications for business productivity. Operational activities are often likened to a caretaking role, although the emphasis here is still on management rather than the servicing trades.

At all levels facility management underpins corporate goals and is an essential component for their achievement. Properly designed and managed facilities lead to productive practices that add value to an organization's bottom line. Workplaces are analogous to ecosystems, so the term 'workplace ecology' reflects the health and prosperity of all its inhabitants. In this context workplace ecology is a key to business continuity and success.

This first part deals with the critical topic of corporate goals. Chapter 1 discusses the concept of workplace ecology in more detail and explores its relationship with broader business drivers. Chapter 2 looks at ways in which worker productivity can be improved through a better understanding of the impact of workplace design on performance. Chapter 3 explores the role strategic facility management plays in achievement of organizational goals and introduces the strategic facility plan.

Implicit in this discussion is the need for identification and analysis of key business drivers and the benchmarking of performance with other organizations involved in similar (often competing) activities. Performance trends from one year to the next are also critical to ensure that facility management is effective and seen to be so. Often tracking measures are goal-orientated and expressed in terms of overall profit and tangible stakeholder benefits.

Workplace ecology

1.1 Introduction

In business, decisions are made in the context of improving quality, reducing costs and minimizing risk. Organizations must operate in an environment of increasing competition, globalization and technological development. The workplace, whether it be an office, factory, hospital, hotel, theatre, TV studio, power station or space station, needs to enable effective performance of its core-business functions.

All workplaces comprise facilities that must be managed. At one end of the spectrum this involves cost monitoring and maintenance, and at the other end involves strategic alliances and integrated business processes. A 'healthy' workplace is one that performs at its optimum. Workplace ecology is the study of workplace performance, understanding the link between human resources (people) and physical resources (facilities). Productivity is this link. Measurement of productivity is the best means of quantifying improvements in workplace ecology.

1.2 Facilities

Workplace ecology is concerned with enabling effective business practice through the provision of appropriate facilities. It embraces issues of organizational reform, worker productivity, information technology, sustainability and occupational health and safety. Facilities need to be designed which are flexible enough to adapt to changes in the above issues without unnecessary waste and cost.

Bernard Williams Associates (1999) defines facilities as 'the premises and services required to accommodate and facilitate business activity'. In other words, facilities are the infrastructure that supports business. This is a wide definition and is intended to cover not just land and buildings but other infrastructure such as telecommunications, equipment, furniture, security, childcare, catering, stationery, transport and satellite work environments (home, car, client office, airport lounge, hotel lobby, etc.). Facilities should not be seen as an overhead, but rather as an integral part of an 'ecosystem' necessary to enable people to perform at their best. The most valuable asset of the organization is its people. The two main support activities for any organization are therefore human resource management and facility management.

1.3 Human resource management

Human resource management is responsible for aligning the critical functions of recruiting, staffing, training and development, and performance management so that they support business strategy. This means that the role of human resources is to work as a strategic partner within the organization to maximize the value of its employees (Bobrow, 1998). Building from the organization's strategic tactics, human resources can plan how its major functions can help the organization carry out its strategy.

To maximize the value of an organization's employees, one of the first steps is to develop a list of the core competencies that individuals need to possess for different positions within the company. This step is a building block for all of the other human resource management activities. Once human resources understands what the core competencies are for a job, it is in a position to actively seek people for the position, and then continue to build upon the core competencies through career development and promotion. Hiring decisions based on these competencies ensures that job openings are filled by the best possible candidates. Human resources should try to attract candidates who have a strong sense of purpose or value system aligned with the company's mission.

Training is a critical component of human resource management, and has the greatest value when it is improving the performance of existing employees to match changing business needs and anticipates future trends. Of all human resource management functions, performance management offers the largest potential return for an organization. Performance management systems should also be linked to organizational objectives. A system that truly rewards excellent performance should:

1. Contain meaningful measures and standards within a specific time frame.
2. Be flexible with changing business needs.
3. Be linked to training and development opportunities.
4. Be related to organizational objectives. (Bobrow, 1998)

By having these measures based on the business strategy, success can be measured accurately and valuable employees can be justly rewarded. If these performance measures are not reached, they should be linked to training and development opportunities so employee performance can be improved.

Human resource management and facility management functions converge in many areas. Finding temporary accommodation during renovation works, setting up new staff so they can begin working productively as soon as possible, communication of complaints about indoor air quality (IAQ), and workspace design are some examples. Co-operation between human resource management and facility management is important, not just because it may solve a specific problem, but also when it comes to tackling larger and more nebulous issues like productivity and change in the corporate culture.

Facility managers have to work more closely with the human resources department to develop policies that reflect changing worker dynamics (McMorrow, 1995). Creating a more efficient, effective and creative workplace to better support changing business practices involves integrating office design, technology, working practices and workplace culture (Mackay and Maxwell, 1997). New technology has enabled the initiative to take place, encouraging a working culture that is status-free, creative, productive, based on knowledge sharing and effective communication. Employees are increasingly dictating what they want from a job, so organizations need to think imaginatively about attracting, retaining and rewarding people. This is where human resource management and facility management can work together to create an innovative approach to integrating people and technology in a way that is beneficial to everyone and to the organization.

There have been dramatic changes in how people occupy space, with 20% more people working in 30% less space (McMorrow, 1996). The focus of alternative work strategies should be on employee productivity and the ability to function in alternative environments, with rent savings as a bonus. The social re-engineering of the workplace involves merging electronic information access with the physical requirements. As people are now capable of working virtually anywhere, human resources has a role to play in making the location of work seamless to the customer. Between the physical and technological investments, human resource management has to help employees mesh with technology and adopt ways of working that empower them more than in the traditional workplace. The human resources department must work with the facility department and designers to ensure that all workers understand all implementations and how they affect each individual. Human resources may help to conduct a needs-assessment survey to identify the needs of each business unit (Anonymous, 1998).

Different issues are involved with corporate building campuses (Leonard, 1999). These self-contained campuses contain day care facilities, cafeterias, convenience stores, automatic teller machines, concierge services, fitness centres, etc. Major corporations (particularly those in the high-tech fields) often centralize their operations in the suburbs, where real estate prices are lower and the land is available, away from the congestion and pollution of the city. These complexes are convenient for employees, allow them to better balance work/life issues, and they foster better teamwork and co-operation among company divisions. Human resources should be aware of issues such as relocation costs, day care policies and procedures, liabilities and the costs of the amenities versus worker retention and productivity.

1.4 Facility management

Facility management is a relatively new discipline. It has grown from the building superintendent or caretaker role to one of significant importance and influence. Nevertheless, the facility manager is not normally part of senior management, but is increasingly required to provide strategic advice that can align an organization's infrastructure to support its core business activities.

Alexander (1998) states that facilities are commonly an organization's second largest expense and can account for up to 15% of turnover, and typically is the largest item on the balance sheet. Although the traditional pressure has always been to reduce costs, decisions are now being taken in the light of maximizing value and competitive advantage. The modern role of the facility manager is a key strategist concerning all matters of infrastructure, including the important area of information technology. Another emerging driving factor is sustainability and environmental performance. Integration of facility information using computerized management systems is now commonplace.

Due to the central role in supporting the business, the facility management function is a key player in the continuous strive towards innovation, performance enhancement and cost reduction. The facility management function is that of a crucial service supplier to many internal customers which can, in turn, significantly influence the level of quality and service delivered to the external, paying client. Today's facility manager must possess knowledge, skills and performance levels far greater and more challenging than before. The wider scope of responsibilities given to the position means that facility managers must be:

1. 'Big picture' oriented.
2. Knowledgeable about both facilities and customers.
3. Adept at financial analysis.
4. Able to properly measure facility performance.
5. Able to objectively evaluate options to best serve facility customer needs.
6. Able to flexibly evaluate and change processes.
7. A marketer, communicator, and positive advocate for facility management. (Slember, 1998)

Core competencies for facility managers are variously defined by professional associations worldwide. These can be broadly classified as comprising expertise in as many of the following areas as possible:

1. Strategic management.
2. Space management.
3. Information management.
4. Risk management.
5. Human resource management.
6. Financial management.
7. Operations and maintenance management.
8. Real estate management.
9. Project management.
10. Conflict management.
11. Asset management.
12. Quality management.

Facility management operates at three levels: strategic, tactical and operational. Today the discipline is very much focused at the strategic level, and pushing to take a more influential role in organizational direction and decision-making. One vehicle that may assist this move is the effective use of information technology and systems that can be woven through the fabric of the entire organization.

1.5 Enterprise resource planning

Enterprise resource planning (ERP) is at the cutting edge of the facility management discipline. It essentially involves the merging of human resources and facilities through interconnected databases and management practices. Productivity is at the heart of ERP, and is attained by enabling people to achieve their best by providing an environment conducive to the tasks they are required to perform. Through proper management, increases in productivity will lead to great profitability for the organization and greater employee satisfaction. This in turn enables quality staff to be 'gained and retained'.

For the workplace to function as a strategic element in the enterprise, it must be recognized that its success depends on the dynamic interrelationship and mutual reinforcement of spatial, organizational, financial and technological issues (McMorrow, 1999). It is important to achieve an active coherence between the work and these aspects of the workplace. ERP allows this through the integration of business information with its processes. Its purpose is to promote a holistic view of a organization's operation and create a master schedule of all its components, stock, employees and equipment across an enterprise in order to increase the efficiency of its performance and improve customer service (Niven, 1999). Its scope should cover production, inventory, accounting, customer service, sales and follow-up activities. The key is optimization of an entire process, rather than of individual, constituent parts. Due to the global and volatile nature of the market, business needs to be planned on a weekly or daily basis, rather than monthly, and therefore it is critical to synchronize the demand all the way along the entire supply chain. ERP focuses on process, from beginning to end, and on all the points of interface and interaction of all operations.

ERP software applications integrate data from all dimensions of an enterprise. ERP implementation usually results in the separation of the front office from the back office with support functions moving out to less expensive real estate, and the elimination of a number of administrative positions. ERP affects key business processes, changing the way people work and the logistics of physical adjacency and co-location arrangements (McMorrow, 1999). ERP systems usually target the mainstream parts of a business, such as the finance, production and logistics.

Computerized maintenance management systems (CMMS) systems are often integrated with ERP systems which enables CMMS to go beyond individual equipment, tools and maintenance procedures to fit more into the big picture of an organization. Integrating maintenance systems to ERP and other business systems creates 'synergy' (Gould, 1998) and allows for better co-ordination of predictive and preventive maintenance with business operation needs. When a CMMS package is part of an ERP system, integration is normally extensive and flexible (Cooper, 1998).

ERP systems integrate business by consolidating information across different parts with a shared online database. ERP has the capability to streamline a business process, but also has the potential to bankrupt it, particularly if it is regarded as just another IT project or cost saving tool (Niven, 1999). It needs to be well integrated, with the decision to implement it a strategic one. According to Gould (1998), ERP may become superseded by advanced planning and scheduling (APS) routines. These will replace the conventional material requirement routines found in classical

ERP systems, particularly in production planning and manufacturing scheduling systems.

References and bibliography

Alexander, K. (1998). 'Facilities Management: A Strategic Framework'. In *Facilities Management: Theory and Practice* (K. Alexander, ed.), pp. 2-11, E. & F.N. Spon.
Anonymous (1998). 'Even Executives are Losing their Offices'. *HR Magazine*, 43(4), March, p. 77.
Atkin, B. and Brooks, A. (2000). *Total Facilities Management*. Blackwell Science.
Barrett, P. (1995). *Facilities Management: Towards Best Practice*. Blackwell Science.
Bernard Williams Associates (1999). *Facilities Economics*. Building Economics Bureau Limited.
Bobrow, W. (1998). 'Is your HR Department in Shape to Support your Business Strategies?' *Bobbin*, 39(8), April, pp. 64-68.
Cooper, C. (1998). 'An Integrated System for Mill Maintenance'. *Pulp and Paper International*, 40(12), December, pp. 19-21.
Cotts, D. G. (1999). *The Facility Management Handbook* (2nd edition). AMACOM.
Gould, L. S. (1998). 'Keeping Up, Running and Profitable with CMMSs'. *Automotive Manufacturing and Production*, 110(8), August, pp. 68-71.
Leonard, S. (1999). 'Is a Corporate Campus in your Future?'. *HR Magazine*, 44(10), October, p. 215.
Mackay, M. and Maxwell, G. (1997). 'Future Shop', *People Management*, 3(20), October, pp. 44-47.
McGregor, W. and Then, D. (1999). *Facilities Management and the Business of Space*. Arnold.
McMorrow, E. (1995). 'Teaming Unites FM and HR'. *Facilities Design and Management*, 14(9), September, p. 9.
McMorrow, E. (1996). 'Real Estate Nods to HR'. *Facilities Design and Management*, 15(12), December, p. 9.
McMorrow, E. (1999). 'CIR, ERP are in your Future'. *Facilities Design and Management*, 18(6), June, p. 9.
Niven, L. (1999). 'Integrating Business Strategy and Enterprise Resources Planning'. Thesis, University of Technology, Sydney.
Park, A. (1994). *Facilities Management: An Explanation*. Macmillan.
Rondeau, E. P., Brown, R. K. and Lapides, P. D. (1995). *Facility Management*. John Wiley & Sons.
Slember, R. J. (1998). 'Measuring Facilities Management Results'. *Office Life Canada*, December. <http://www.facilitiesnet.com/fn/NS/NS3o8la.html>
Tompkins, J. A. (1996). *Facilities Planning* (2nd edition). John Wiley & Sons.

Worker productivity

2.1 Introduction

If facilities are the infrastructure that supports business, then worker productivity is a key objective of the facility manager. Facilities must enhance productivity through provision of appropriate workspace, technology and physical environment. But even more so, facilities must enable organizational flexibility, task-orientated activities and user customization.

Productivity is crucial to survival in today's business environment. Diminished profits, increased local and global competition, and diversity in the range of skills required are driving organizations to look for methods to make themselves more productive, sometimes with fewer people and/or in less space. 'Corporations are increasingly judging facility executives on their ability to provide office workspaces that make it possible for employees to be more productive' (Sullivan, 1996). Apart from increased earnings, increased productivity can result in energy savings, reduced healthcare costs, reduced costs due to absenteeism, staff churn and real estate, hence improving profits further. Identity and image are also very important in the market, with the appearance of the office space conveying a message about the company.

Surveys suggest (Anonymous, 1997b) that many business decision-makers recognize the importance of their companies to appear flexible, adaptable and forward thinking. Presenting a positive corporate image through a productive environment and employee satisfaction speaks volumes. The design of workspaces is important in attracting and retaining the types of employee best suited to the corporate culture. The challenge for facility managers is to find improvements that will produce a better work environment with a return of higher productivity.

2.2 Productivity and work environments

The question that many businesses wrestle with is whether to cut costs or support productivity – and some try to do both simultaneously. Most often, however, the two doors to workspaces, and creating a separation from service areas, is also critical. Another non-tangible strategy is allowing for flexi-time (Merisalo, 1995) so that employees can work earlier or later if they wish. Implementations such as these can result in lower noise levels, higher worker satisfaction and improved profits (Pascarelli, 1994).

Apart from addressing physiological and environmental factors that impair the performance of workers, other commodities such as technical and space requirements need to be considered. These need to be flexible to accommodate changes in organizations so that productivity is not hampered. For instance, cabling and wiring needs to be modular, extendable and movable. The facility manager has to have an eye towards future upgrades in computers and other electronic equipment.

The items discussed above are more or less housekeeping issues and while not ignoring them, the facility manager must also pay great attention to the new productive ways that people are working. Employees are increasingly working together with other people in multiple work settings rather than working individually. In the last ten years or so the amount of time people have spent working individually has reduced from 70% to 25–30% of the day (Scott-Berry, 1996a). Therefore the facility manager must respond to this by allocating space to encourage group interaction and provide shared areas with appropriate facilities for group work. Office space must be flexible enough to accommodate staff when they are working together as a team and when they are working alone. An office space strategy designed to ensure easy contact and communication between team members (physically or electronically) and provide access to other vital resources will encourage co-operation, collaboration and effective teamwork (Vergetis-Lundin, 1998).

2.4 The role of the facility manager

Research results suggest that high quality buildings can increase employee productivity by 10–20% (Anonymous, 1997a). However, research also shows that the majority of corporate strategic plans do not include real estate and workplace objectives, even though the dividends of enhancing the building performance could be quite high (Vergetis-Lundin, 1997). This could be partially due to the facility manager's place in the typical organizational hierarchy. Most facility managers need to become more business and corporate goal-oriented and be part of the planning and decision-making process. To maximize the contribution of facilities to a company's performance, the facility manager must identify the objectives and priorities of corporate strategy. In fact, the facility manager must become a business strategist and truly understand the company in order to provide the best solutions.

As a step towards creating an integrated approach to corporate infrastructure decision-making, the facility manager should co-ordinate with end users, human resources, information systems, senior management and CEOs and ensure that all the goals and directives are consistent. The effectiveness of a strategy is far greater when part of an integrated masterplan rather than in isolation (Keppler, 1997). Facilities is one component of a system required to support workplace changes, and therefore collaboration with senior management and other departments will produce a fully integrated programme with financial and psychological backing needed to transform the work environment. All contributors have crucial input when it comes to evaluating the environment and how it can increase their productivity. Working as a team in this way leads to far better results as there is a larger pool of knowledge from which to base decisions.

Corporate attitude toward the impact of facilities on production is gradually changing and top management is realizing that human resources are its greatest asset and must be properly supported (Tatum, 1998). Upper management has to be made aware of the long-term benefit of facility-related investments. The return comes from employee satisfaction and increased productivity, reduced staff turnover, reduced absenteeism, fewer customer complaints and better handling of customer needs. A possible line of reasoning to explore is to make the organization aware that space is not merely a cost but an asset that can be leveraged. Vergetis-Lundin (1997) states that facility managers 'are looking specifically at the size of the workspace, where it should be located, and what will make it comfortable and productive as a way to enhance the workplace and improve business results'. The buying of acquisitions, leasing, disposition and design should be done in communication with human resource management so that new employees can be productive as soon as they start.

The initial and simplest step towards starting a programme of improving productivity in the workplace is to identify and remove obstacles to worker effectiveness. An excellent way to understand the impact of the facilities on employee performance is to listen to or ask the employees. For instance, one of the most frequent complaints from workers is that offices are either too hot or too cold (Vergetis-Lundin, 1998). A survey by ASID indicated that employees believed they would be more productive if they had fewer noise distractions in the office, more storage and filing space, larger workspaces and more open space for teaming (Anonymous, 1996). By carrying out a survey within an individual company, the facility manager will find out how the employees work and obtain suggestions as to how they can work better. Workers' solutions may be less expensive than those of management (Scott-Berry, 1996b) and staff participation will minimize resistance to change (Anonymous, 1997b) and will engender more respect for the workplace from the employees.

References and bibliography

Alexander, K. (1996). *Facilities Management: Theory and Practice*. E. & F.N. Spon.

Anonymous (1996). 'A Worker's Perspective: Design Factors Likely to Promote Productivity, says ASID Survey'. *Facilities Design and Management*, 15(2), February, p. 16.

Anonymous (1997a). 'Top Management Needs to Hear the FM Message'. *Facilities Design and Management*, 16(4), April, p. 9.

Anonymous (1997b). 'Design Does Make a Difference in Worker Productivity'. *Facilities Design and Management*, 16(9), September, p. 16.

Anonymous (1999). 'A Well-designed Office is a Productive Office'. *Facilities Design and Management*, 18(1), January, p. 13.

Atkin, B. and Brooks, A. (2000). *Total Facilities Management*. Blackwell Science.

Barrett, P. (1995). *Facilities Management: Towards Best Practice*. Blackwell Science.

Cotts, D. G. (1999). *The Facility Management Handbook* (2nd edition). AMACOM.

Damiani, A. S. (1998). *Moving up the Organization in Facilities Management: Proven Strategies to Increase Productivity in your Workforce*. Scitech Publishing.

Green, G. M. and Baker, F. (1991). *Work, Health, and Productivity*. Oxford University Press.

Guzzo, R. A. and Dickson, M. W. (1996). 'Teams in Organizations: Recent Research on Performance and Effectiveness'. *Annual Review of Psychology*, 47, pp. 307-38.

Hansen, Wayne (1995). 'The IAQ Challenge to Facility Management: Healthy buildings through affordable indoor air quality programmes'. *Facilities*, 13(12), November, pp. 12-20.

Hesselbein, F., Goldsmith, M. and Beckhard, R. (1997). *The Organization of the Future*. Jossey-Bass.

Keppler, K. (1997). 'Bridging the Gap between Work and Workplace: Careful selection of an alternative workspace strategy is a key part of the process of transforming an organisation'. <http://www.facilitiesnet.com/fn/NS/NS3b7ha.html>.

McGregor, W. and Then, D. (1999). *Facilities Management and the Business of Space*. Arnold.

Merisalo, L. (1995). 'Acoustics: Factoring sound, or the lack thereof, into the space-planning/productivity mix'.
<http://www.facilitiesnet.com/fn/NS/NS3pdaco.html>.

Park, A. (1994). *Facilities Management: An Explanation*. Macmillan.

Pascarelli, E. (1994). *Repetitive Strain Injury: A Computer User's Guide*. John Wiley & Sons.

Rondeau, E. P., Brown, R. K. and Lapides, P. D. (1995). *Facility Management*. John Wiley & Sons.

Rosenblatt, B. (1995). *New Changes in the Office Work Environment: Toward Integrating Architecture, OD, and Information Systems Paradigms*. Ablex Publishing Corporation.

Salvendy, G. (1997). *Handbook of Human Factors and Ergonomics* (2nd edition). John Wiley & Sons.

Schulze, P. C. (1999). *Measures of Environmental Performance and Ecosystem Condition*. National Academy Press, Washington.

Scott-Berry, E. (1996a). 'Workplace Planning Strategies: Productivity – effective workspaces remove obstacles to efficiency'. *Building Operating Management*.
<http://www.facilitiesnet.com/fn/NS/NS3b86f.html>

Scott-Berry, E. (1996b). 'Workplace Planning Strategies: Flexibility – new work styles need adaptable workspaces'. *Building Operating Management*, August.
<http://www.facilitiesnet.com/fn/NS/NS3b86e.html>

Stevens, D. (1997). *Strategic Thinking: Success Secrets of Big Business Projects*. McGraw-Hill.

Sullivan, E. (1995). 'Does Office Design Hinder Employee Productivity? The answer is yes, according to many corporations. Here's what they're doing about it'. Trade Press Publishing Corporation. <http://www.facilitiesnet.com/fn/NS/NS3bh5a.html>

Sullivan, E. (1996). 'Pay Attention to Acoustics: Most offices don't; employees pay the price'. *Building Operating Management*, August. <http://www.facilitiesnet.com/fn/NS/NS2bmaug.html>

Sweeney, D.B. (1998). 'Foreword'. In *Facilities Management: Theory and Practice* (K. Alexander, ed.), E. & F.N. Spon.

Tatum, R. (1997). 'Tackling Workplace Problems: The physical environment can boost or limit productivity in seven key areas'. *Building Operating Management*. <http://www.facilitiesnet.com/fn/NS/NS3b7ec.html>

Tatum, R. (1998). 'Stepping Stones to Productivity: It's impossible to deny that building technologies have an impact on employee performance – the hard questions are how – and how much?'. *Building Operating Management*, May.

Tompkins, J. A. (1996). *Facilities Planning* (2nd edition). John Wiley & Sons.

van Delinder, T. (1997). 'Creating a Progressive Office Environment'. *Facilities Management Journal*, May/June, pp. 18-22.

Vergetis-Lundin, B. L. (1996). 'Path to Productivity: Quality lighting can offer productivity improvements and energy savings'. *Building Operating Management*, August. <http://www.facilitiesnet.com/fn/NS/NS2bmaug.html>

Vergetis-Lundin, B. L. (1997). 'Integrating People, Places and Processes: Facility executives play an important role in company-wide efforts to boost productivity'. *Building Operating Management*, May. <http://www.facilitiesnet.com/fn/NS/NS3b7eb.html>

Vergetis-Lundin, B. L. (1998). 'Designing Productive Office Space: Keep in mind that corporate image, technology and the need for team spaces play a role when building high-performance workspaces'. *Building Operating Management*, August. <http://www.facilitiesnet.com/fn/NS/NS3b8ha.html>

Strategic facility management

3.1 Introduction

A facility should support business goals and corporate culture. To maximize the facility's contribution to company performance, the facility manager must understand corporate priorities and make sure his or her goals and directives are consistent with them. The facility manager should question whether the operation of the workplace reflects the corporate goals. To incorporate real estate and workplace objectives into corporate strategic plans, the facility manager should, at best, become one of a team of decision-makers or, at least, report directly to the company's chief executive officer (CEO). Companies should be made to think of facility management in terms of an integral business process and have CEOs incorporate facility management into their overall strategic and business plans.

Accordingly, facility managers should work closely with organization decision-makers to develop the best facilities model possible to support the business and its operational goals into the future. The tools used by the facility manager should produce valuable information on space, maintenance and capital assets that can be easily incorporated into the decision-maker's overall planning data and used in strategic and business planning processes. This should be a collaborative process in which the facility manager is part of the planning solution.

Strategic facility planning requires a proactive rather than reactive approach to real estate decision-making; it requires forward thinking to anticipate problems that may arise in the future and in turn demands contingency plans to address those problems if they arise.

3.2 Management focus

Facility management can encompass operational, tactical and strategic levels. While all three might be important, it is generally regarded that the real potential for business enhancement lies with strategic facility management.

Operational issues concern day-to-day management of facilities, and focus on the caretaker role which is the foundation of the modern facility manager role. Maintenance and repairs, security, car parking, building services and gardening are examples of operational issues. While undoubtedly important, they are largely of a technical nature and low down in the management hierarchy.

Tactical issues concern customization and planning, and focus on understanding the needs of the organization in terms of facility support and implementing solutions that add value. More than a technical task, it embodies management of processes and co-ordination of effort to deliver facilities which reflect the organization's needs.

Strategic issues go beyond meeting needs; they attempt to predict organizational direction and provide integrated solutions. Facility management is defined here as a high-level management activity, in concert with the strategic planning of the organization, and with access to or participation in key decision-making processes.

The creation of a facility strategic planning function opens a new role for facility managers. No longer simply a provider and manager of productive work environments, the facility manager is now an active participant in planning the future of the organization. If the goal is to be a leading-edge firm in a business environment that has become global, where technology and competitive alliances are the norm, a facility manager needs to use intellectual leadership, perseverance and understanding of competition in order to provide fundamental new customer benefits. A facility manager therefore must understand that influence comes from:

1. World-class core competencies.
2. Knowing how to develop relationships and work with partners.
3. Industry foresight.
4. Knowing what they want to give customers ten years from now. (Hubbard, 1994)

3.3 The strategic facility manager

It is not enough to be a real estate strategist. Strategic facility planning requires a business background and quantitative skills. The facility manager needs to be a business strategist and truly understand the business groups he or she is working with in order to provide them with the best service. In order to become more strategic, facility managers should not become slaves to daily operations, but rather think and act in business terms and be bottom-line oriented. They should keep abreast of industry, market and technology trends.

It is an advantage to think of the users of facilities as customers, as this will foster better working relationships and partnerships with them. Facility managers must learn to think like their clients in order to get a clear picture of the broader business goals and expand the scope of their charge to meet them. Determining how each of the business sections comprising the organization values their own success gives the facility manager a template against which to measure the value of the services provided. Further, understanding how the sections and top management make decisions on new initiatives gives facility managers the ability to anticipate the tools needed to succeed.

Forward-thinking facility managers will keep track of not just traditional 'facili-

vill focus on other
pport services. By
drivers, the facil-
de influential pre-
tability (Bell and

ng include:

ten years.
te for future com-

nation of staff and

h senior manage-

age.
sibilities and out-

ticipate future cir-

nanagement.
stomer (Hubbard,

ve and maintain a
nd with business
increase revenue.

3.4 Identifying corporate objectives

High performing organizations strive for maximum quality and productivity simultaneously by being customer focused, people oriented and by value-adding their primary processes while keeping unit operation costs down. These organizations develop their strategies by taking a long-term approach to business growth. Within this environment, the facility manager's role is to provide services that enhance the organization's ability to flourish in a volatile commercial climate. Logically then, facility management shares the same challenges as the organization and should follow the same goals and *modus operandi* (Tranfield and Akhlaghi, 1995).

To make sure the customer gets the best solution and to achieve an integrated, co-ordinated approach to corporate infrastructure, facility managers should work cross-functionally with end users, human resource managers and senior management. This promotes faster communication and better results due to a more knowledgeable decision-making team. Information and strategy exchange is essential in producing a useful strategic facilities plan. When compiling the plan to embody corporate policy, the facility manager should seek information from such sources as the operating plans for the organization's strategic business units, general policies and business plan, and historical data of the organization and its interfaces (Brown et al., 1994).

So that facility managers are not caught off-guard by unexpected changes, they should become aware, or better still, become part of the business decision-making process responsible for that change. For example, with changes in requirements for space and infrastructure to support staff shifts, the facility manager can anticipate demand instead of merely reacting to it by realizing that churn (staff reassignment) is the direct result of the need to meet the demands of the market. By keeping an eye on market trends and being aware of the needs of the organization to keep astride of these trends, the facility manager may anticipate changes in business goals and approach decision-makers early in the process. The facility manager is then prepared to support that effort by anticipating staff need for appropriate workspace and access to technology (while containing expenses). The final solution should meet the business need and increase efficiency (Bell and Keppler, 1997).

Each section of a business has its own success measures that correspond to its function and outcomes (for example, sales and marketing focus on those elements that stimulate demand and customer satisfaction; operational information focuses on response time, cycle time and improved business processes; human resources and real estate services are directed toward recruiting and retaining staff and creating an environment to improve productivity and performance). Adopting the organization's success measures will optimize the management of facilities. This creates the starting point for the co-ordination or creation of the components necessary for achievement of the organization's goals. As the facility manager moves toward a more influential corporate role, the understanding and integration of the measures that drive the business becomes more important to long-range planning processes.

3.5 Strategic facility planning

A strategic facility plan (SFP) is a term used to describe the consolidation of modern facility management activities. Much information necessary to anticipate facility requirements is regularly gathered and analysed by developers of business strategy, product planning and competitive analysis. Communication links with finance, personnel and organizational planning groups allow facility requirements to be identified and alternative strategies discussed, as business plans are being developed. An appropriate proactive and flexible SFP results from a joint effort of all affected organizations.

The steps required to create a strategic, long-range (three to ten years) and short-range (two to three years) SFP are as follows:

1. Forecast the facility implications of any new future business scenarios/objectives (for example, gross space, location and amenity requirements).
2. Compare forecasts to existing owned/leased facility resources.
3. Construct broad facility solutions and budget direction alternatives to support business plan objectives.
4. Annually update the plan and budget to support improved financial forecasting to decision-makers/senior management.

The long-term SFP should also:

5. Support the functions of short-term facility master (operations-level) planning through regular sharing of long-range strategic facilities plans and budgets.

Additionally, the short-term SFP requires:

6. Production of a prioritized annual facility project plan and budget for sign-off to, and implementation by, detailed planning-design.
7. Regular co-ordination of all ongoing short-term facility plans and budgets with detailed design/construction management.

The formation of a truly strategic long-range SFP requires the support of senior management, which means that the financial and organizational benefits need to be clearly identified, communicated, understood and agreed.

3.6 Potential problems and issues

Problems encountered when establishing strategic facility planning are related to forecasting accuracy, particularly future space capacity requirements. Forecasting even gross space capacity need requires some understanding and agreement regarding the amount of space utilized by at least the normal space consuming categories (i.e., people, storage and technology). Before facility implication forecasts can be made, the entire strategic business and facility planning team needs to analyse and agree upon these and other important points.

The objectives and information of a strategic corporate business plan, such as improving cost performance, products/services and market share, may be controversial and could result in serious political, business, employee, union and/or other consequences. The facility manager needs to be sure that sensitive objectives and data are not compromised by limiting business-planning data to that which is necessary to identify the required facility implication projections.

In order to integrate a strategic business plan with an SFP, an understanding of the psychological and sociological needs of the organization's employees must be gained. This comprises their values, issues, processes and relationships. Company culture is a matrix of several factors, such as:

1. The attitude of the founder.
2. Belief systems, company values, expectations and decision-making.
3. Normative behaviours – how decisions are made and crises are handled.
4. Formal organization reporting hierarchy.
5. Informal organization reporting ('the grapevine').
6. Information systems and technology (what the employees and public know about the organization).
7. Organizational culture. (Giarrusso, 1997)

For a successful SFP, all of these factors must be considered. One way to achieve this is to gain insight into what the employees think needs to be changed and how they would prefer it to be changed.

It is important to recognize that in the knowledge-based enterprise, information is power. Old determinants of competitive advantage like size and market share are being surpassed by intelligence (people), connectivity (technology) and speed – the intellectual capital of the organization (Bell and Keppler, 1997). Enabling the synergy between people and technology, the facility manager has an unusual opportunity to capture a significant component of the intellectual capital of the enterprise. The greatest opportunity in this is not just to collect the data but to translate information into knowledge and relay it back to the business units within the organization. The facility manager's role can then evolve into one which integrates this information to enable the strategic planning process to be more effective and more relevant, which in turn will enable the business enterprise to be more successful.

Facility management, technology and parts of human resources can be combined into one service unit to create a more integrated service delivery operation. This benefits the business units and company as a whole by minimizing the numbers of contacts required to obtain full support services and, at the same time, reducing the number of steps required for delivery. For example, by linking the human resources personnel tracking system, the corporate real estate database and the information management system, a more effective platform for tracking and, ultimately, forecasting occupancy requirements is created.

The facility manager can take control of the support services process and be accepted as a strategic 'business' expert, particularly with the right technology required to create the necessary tools and networks. It is up to the facility manager to step out of the 'space box' and leverage their knowledge of product, services and process.

3.7 Knowledge requirements

Facility managers need to have a diverse educational background, since the discipline has grown to be a composition of a large range of competencies. While not needing to be an expert in any particular field, the facility manager should have a good understanding of the key areas of management related to the built environment.

Zulkeski (1997) describes two types of facility managers: the caretaker and the change agent. While both perform necessary tasks, the majority of caretaking functions can be outsourced, while the change agent functions represent value-added facility management. People, process/technology and place in an organizational context are all undergoing radical change and the facility manager who is a change agent is best equipped to deal with these shifts. There will be significant changes in organizations, driven by a generally and technically more informed and experienced population, creating major shifts in the demographics of the labour force on a global basis.

References and bibliography

Alexander, K. (1996). *Facilities Management: Theory and Practice*. E. & F.N. Spon.
Atkin, B. and Brooks, A. (2000). *Total Facilities Management*. Blackwell Science.
Barrett, P. (1995). *Facilities Management: Towards Best Practice*. Blackwell Science.
Bell, S. and Keppler, K. (1997). 'Strategy and Integration: The FM's broader charge'. *Facilities Design and Management*, 16(10), October, pp. 59-61.
Brown, R. K., Lapides, P. D. and Rondeau, E. P. (1994). 'Corporate Policy as Part of RE/FM Planning'. *Facilities Design and Management*, 13(7), July, pp. 50-53.
Cotts, D. G. (1999). *The Facility Management Handbook* (2nd edition). AMACOM.
Damiani, A. S. (1998). *Moving up the Organization in Facilities Management: Proven Strategies to Increase Productivity in your Workforce*. Scitech Publishing.
Giarrusso, T. W. (1997). 'Marketing a New Society'. *Facilities Design and Management*, 16(5), May, pp. 46-51.
Guzzo, R. A. and Dickson, M. W. (1996). 'Teams in Organizations: Recent research on performance and effectiveness'. *Annual Review of Psychology*, 47, pp. 307-38.
Henderson-Paradis, R. (1996). 'The Strategic Planning Process and the CEO: Facilities managers need the tools, technology and new processes to provide top management with important information'. *AFE Facilities Engineering Journal*.
<http://www.facilitiesnet.com/fn/NS/ NS3a96e.html>
Hesselbein, F., Goldsmith, M. and Beckhard, R. (1997). *The Organization of the Future*. Jossey-Bass.
Hubbard, G. M. (1994). 'Why Should You Become More Strategic and Less Tactical?'. *Facilities Design and Management*, 13(2), February, pp. 50-53.
McGregor, W. and Then, D. (1999). *Facilities Management and the Business of Space*. Arnold.
Park, A. (1994). *Facilities Management: An Explanation*. Macmillan.
Rondeau, E. P., Brown, R. K. and Lapides, P. D. (1995). *Facility Management*. John Wiley & Sons.
Stevens, D. (1997). *Strategic Thinking: Success Secrets of Big Business Projects*. McGraw-Hill.
Tompkins, J. A. (1996). *Facilities Planning* (2nd edition). John Wiley and Sons.
Tranfield, D. and Akhlaghi, F. (1995). 'Performance Measures: Relating facilities to business indicators'. *Facilities*, 13(3), March, pp. 6-14.
van Delinder, T. (1997). 'Creating a Progressive Office Environment'. *Facilities Management Journal*, May/June, pp. 18-22.
Vergetis-Lundin, B. L. (1997). 'Integrating People, Places and Processes: Facility executives play an important role in company-wide efforts to boost productivity'. *Building Operating Management*.
<http://www.facilitiesnet.com/fn/NS/NS3b7eb.html>
Zulkeski, A. G. (1997). 'The FM as Change Agent'. *Facilities and Design Management*, 16(1), January, p. 64.

Part 2

Functional plans

Planning is a vital component of any management function, and facilities are no exception. However, the diversity of external and internal factors that can affect future plans is impossible to predict and, needless to say, difficult to manage. Nevertheless short-, medium- and long-term plans must be drawn, reviewed and changed on a continual basis. Without meaningful plans there is no way to compare performance with expectations (as there are no clear expectations) and therefore areas of potential improvement cannot be identified.

There are many types of functional plan that a facility manager might prepare, ranging from financial plans to those involved with specific areas of activity, such as inventories, maintenance, energy usage and the like. Functional plans are the generic names given to business documents that identify methods to deliver corporate goals, which in the case of facility management are infrastructure-related. They are essentially action statements communicated in written form for the information of organization members.

Floor space and its efficient usage are one important instance of functional plans. Space planning enables available physical resources to be allocated to ensure that business processes are properly supported. Too little space may result in an inability to function to an appropriate standard, while too much space will carry cost penalties that the organization may not be able to sustain. Finding the right balance involves working closely with users and understanding their processes and workflow patterns. This includes, but is not limited to, the relationship between different business activities and inefficiencies that might arise due to relocation, inferior communication systems or technology bottlenecks.

Understanding users' needs is a primary role of a facility manager, and comprises collection of facts, opinion, judgement, knowledge of future directions and industry benchmarks. A needs analysis is undertaken during times of expansion or contraction, and particularly when new facilities are to be acquired, to determine the amount of floor space required and how best to use it. Allowance needs to be made for non-useable space within the total gross area suggested by the sum of individual area calculations.

A learning process is often created so that new decisions are based on historical evidence of what has been done successfully and what, unfortunately, has not. Without some feedback mechanisms it is likely that mistakes will repeat and areas of good performance will not be understood. In the transition from perception to reality the question of financial impact arises continually, creating a situation where the various functional plans are all interconnected.

Individual plans form part of a strategic planning process that links all aspects of the organization's activity to current corporate goals. Strategic functional plans are those that provide input to management decisions at a macro-organizational level. Below these may exist a series of tactical and operational plans as required to effectively manage a large facility portfolio. Collectively they serve an additional purpose as change management tools.

Portfolio management introduces some additional strategic issues such as diversity, consolidation, decentralization and geographical positioning as driven by marketplace demand and expectations of financial return. The larger a portfolio becomes the more complex are the issues, but savings can be made through economies of scale and shared processes and systems. It is vital that cost significant items can be identified and recorded so that trends based on actual performance can be correctly interpreted.

This part focuses on the preparation of plans and the context within which they are viewed. Chapter 4 looks at space planning and methodologies available to reduce floor space without correspondingly reducing quality. Chapter 5 investigates how useful data is collected and employed to make effective decisions on space allocations. Chapter 6 discusses the reapplication of data throughout the organization to support more effective portfolio management.

Effective planning is integral with information management and the routine assessment of risk. But understanding organizational need is a starting point and is the vehicle by which corporate goals are established, managed and eventually realized.

Space planning

4.1 Introduction

Companies are assimilating advances in technology, work processes and social dynamics at a rapid pace. Facility managers should be mindful not to lag behind these advances. The new breed of facility manager realizes that there is more to workspace than square metres, tenant allowances, furnishing, fixtures and equipment – space has the potential to enable positive change in an organization and provide competitive advantage.

Space planning has become a difficult task for facility managers, with companies having to respond to rapid changes in the global market coupled with keeping abreast with technology changes. Due to frequent restructuring moves to make companies more productive, churn rates for many companies have increased and can be as high as 70–150% (Tatum, 1997). Corporate reorganization, shifting employee workspaces, the need for higher productivity and reduced operational costs (Scott-Berry, 1996a) have changed the face of office design. As companies may move several times a year, reassembling and adapting to circumstances, emphasis in space planning has been placed on flexibility that enables quick response to changing organizational needs. Add to this changes in work patterns – with employees spending up to 70% of their time working as teams and less time individually (Scott-Berry, 1996a) – and it can be seen why effective space planning is critical. In establishing a corporate identity, there is a shift away from a corporation's appearance – its furnishings and decorations – to how people do their work and what kinds of tools they have (Scott-Berry, 1996b). Perhaps part of a corporation's identity should be a demonstration of its various business functions.

While space planning is often discussed in the context of corporate 'offices', it in fact applies to all facility infrastructure that involves physical workspaces where people undertake tasks, communicate with others and collaborate in teams. The following discussion focuses on office space but the principles equally apply to other work situations. A critical factor is the need to properly assess and analyse options prior to their selection and implementation. This is the role of the facility manager.

4.2 Space management and strategic planning

Space management is traditionally thought of as the skill of maximizing the value of existing space and minimizing the need for new space, and is usually adopted when the outlay required to upkeep facilities is predicted to cost more than the capital resources available in the future. The primary objectives of space management are:

1. Establishing guidelines and procedures for equitable distribution of space to all users based on actual need.
2. Setting parameters for objective evaluation of space use.
3. Ensuring efficient use of space.
4. Establishing a budget and timetable for capital outlay for regular renewal and replacement of facilities. (Henderson-Paradis, 1996)

An added dimension must be given to the objectives to accommodate the way people and companies function today. Issues that the facility manager should consider when planning space include the need to:

1. Balance productivity, job satisfaction, occupancy and capital expenditure.
2. Accommodate new ways of working.
3. Provide for flexibility and future change.
4. Protect privacy and promote teamwork as functions require. (Keppler, 1997)

Thinking in terms of an overall business process and incorporating facility management into the company's overall strategic and business plans may be a radical change in the way most facilities operate. But its importance cannot be overstressed as facilities are often the largest asset of an organization and its proper management in an integrated way will have a great impact on the ability to strategically plan for efficient operation and profitability into the future (Henderson-Paradis, 1996). The facility manager should produce information on space, maintenance and capital assets that can easily be incorporated into an administrator's overall planning data.

In order to plan a space effectively, the facility manager needs to clearly define the scope of the project and the goals to be achieved. The best way to achieve this is to develop a deeper understanding of the company's mission and strategic plan and have discussions with the company decision-makers. Space should be viewed as a set of tools critical to meeting the company's strategic objectives. Space planning should have an effective relationship with the real work of people and link with the broader organizational goals.

Companies should identify their work patterns in order to project both short- and long-term facility implications for space. All organizational areas and facility sites should be analysed, including corporate and subsidiary functions. After completing a broad examination of patterns, a more detailed analysis of specific work processes will enable the facility manager to assess which examples of alternative space would be more appropriate (Ryburg, 1996a). Companies should look at the technologies and organizational behaviours needed to achieve their strategic goals, and then follow through with an analysis of the impact on facilities (Sullivan, 1995).

4.3 Office design and work patterns

In order to understand how the company operates, the facility manager should talk to the employees to ascertain exactly the type of work they perform, how they perform it (alone or with team members) and where the work is done. This information will enable the facility manager to determine the kinds of spaces needed for individuals, teams and larger sections, as well as equipment and storage needs (Vergetis-Lundin, 1998). The best way to improve space efficiency is to rearrange the organization's hierarchy by assigning workspace based on what each person needs to accomplish tasks, i.e., allotting space by what a person does, not who that person is (Scott-Berry, 1996b). Or in other words, 'prioritise space by function rather than hierarchy' (Vergetis-Lundin, 1998). While strategically planning the space to meet the goals of the project team, the needs of individual employees should not be forgotten. As a means to a more successful end, the employees in the planning process should be kept involved by asking for and using their feedback about alternative strategies. Employees will gain maximum advantage from a new design, furniture and/or equipment if they are kept informed, have their input and are given training on how to use them.

Open office concepts have been around for years, primarily as a way to make the building flexible over its lifetime. The open plan developed in the late nineteenth century when office work primarily supported the manufacturing industry. Workspaces were large, open 'bull pen' spaces with tables for several, non-interacting workers. As the industrial age shifted to the information age and business went from national to international, the nature of the office space began to change. Computers transformed office functions from laborious clerical tasks to advanced management duties. Eventually offices have become spaces for activities that require both interaction and concentration to foster communications, creativity and innovation (Considine, 1999). As a result of the new ways corporations are doing business and associated advances in technology, the office has evolved into various alternative arrangements depending largely on how the company and its employees function.

Work patterns are gradually polarizing into mobile work (offsite) at one end and teamwork at the other (Keppler, 1997). Team-based work processes are designed to get more innovative and higher quality products or services to market faster and at lower cost. This demands more open plan, flexible workspaces. Companies are considering a variety of alternative office set-ups in an attempt to get employees closer to their customers, including offices based in cars or customer facilities (Sullivan, 1995). As a result, alternative workplace solutions, including both onsite (team or group space, free address, hotels/motels) and offsite (telecommuting, satellite offices, virtual officing) strategies, that address the range of these changes are among the most successful applications to date (Keppler, 1997).

The design and space planning concept known as activity settings (Scott-Berry, 1996a) is built around a worker's movement throughout the day. The facility manager must design a mix of settings in which people can perform their jobs to create a variety of work areas for privacy, team interaction, research and computer work – catering for increased mobility of the worker within and outside the office. Some employees need a combination of workspaces, including a home office, a workspace adjacent to the team area and a quiet workspace for tasks that require high concentration. Work areas need to be adjusted to accommodate where people are spending most of their time.

The facility manager must be aware of the advantages and potential pitfalls of various office strategies. These are listed in Table 4.1 for comparison.

A survey of where and how employees work will help to determine how to set up the workplace effectively and if alternative officing strategies are best for a company. Identifying the percentage of time each person is in their office will determine if any jobs could be performed at a satellite office. If jobs can be moved out of the office, it is necessary to evaluate whether employees will need to visit the office and how often (Vergetis-Lundin, 1998).

The decision to select an alternative workspace strategy is a key step in the process of transforming an organization and creating room for professional growth and satisfaction at the same time as saving space. For example, Time Inc.'s facility management response to shrinking office space, an expanding staff of roving reporters and home-based writers, a tight budget and little time, was to create a free address work area by replacing three standard 3.7 × 3 metre offices with five mini-workstations each with a computer, telephone with voice-mail programming, and modem line. Sliding Shoji doors allow light into the space and achieve physical and acoustic privacy without claustrophobia in the small space. Some reporters store a mobile pedestal for their days in the office. Thoughtful layouts support the work effort and encourage collaboration with appropriate lighting, work-tables and seating (Keppler, 1997).

To work well, free address office workers may need training in the enabling technologies – computers, electronic mail or other online services, and telephone programming, and have consistent communication and a shared understanding of performance expectations with the employer (Keppler, 1997).

The facility manager needs to provide the employee in the external office with appropriate furniture and office equipment. As such, costs for portable equipment, home office furniture and satellite office equipment with support services must be factored into a space planning analysis.

To accommodate teamworking, office space must be flexible enough to support team members when they are working alone and when working together. To improve teamwork, communication and collaboration, an effective approach is to allow team members to plan their own teaming, or communal, spaces. It is therefore necessary to identify areas where a lot of time is spent and combine several activities in the one area – this can also become a casual meeting space that encourages interaction. For example, a copy room can be turned into a coffee room or a place where several activities can be combined, such as mailing, copying and faxing (Vergetis-Lundin, 1998; Tatum, 1997). The teaming space should be accessible to all team members, and can be either open or enclosed, depending on the needs of the team.

In shared spaces, dedicated storage space is a serious issue, and adequate secure, personal storage space should be provided for employees who visit the office on a regular basis. Be careful not to reduce space too much or create spaces that aren't properly designed with sound ergonomic principles, as this will affect productivity.

Once there is an understanding of the company's business and corporate culture, and how the workers operate and interact, the facility manager will be able to design

Table 4.1 Alternative office strategies

Office Type	Description	Advantages	Disadvantages
Traditional cell	Dedicated to individuals	Privacy; quiet; individual work	Inflexible, doesn't foster interaction and teamwork
Multi-site	Combines separately located dedicated offices and shared team into a single workspace	Laid out to give both individuals and groups maximum location privacy or openness and work arrangement flexibility	Decentralized facility management
Shared space	Shared by two or more; may incorporate work tools, amenities and flexible furniture	Promotes group interaction; can be used as meeting rooms; supports variations in team size, project cycles and duration; gives choice of workspace that best supports each task	Reluctance to give up claim to a particular area; meetings ma disturb non-participants; lack of privacy; lack of secure personal storage space
Hotel space (or 'just-in-time' offices)	Office or workstation is reserved with a phone call	May accommodate more staff without higher facility costs; increased mobility for worker	The space can be easily o booked without adequate planning
Free address (landing site)	Unassigned work areas allotted on a first-come first-served basis (with computers, faxes, programmable phones, and limited storage); 3–5 people per workstation	Unscheduled employees can use whenever needed; better utilization of space; huge real estate savings (30–80%); flexibility and access to required tools; increased mobility for worker	Unscheduled landings ma flict; loss of social fabric of company with absence of dail contact and lack of personalization; lack of co-worker access; lack of access to storage and filing; wayfinding (use of equipment)
Virtual office (satellite offices – telecentres and home)	Anywhere that people can work with the aid of transportable equipment	Space reductions; attractive to employees who face a long commute; flexibility offered by a satellite office centre could boost employee satisfaction	Isolation of stay-at-home w ers; possible inadequate of equipment and software security risks; difficult interactions with central-office employees; conflicts with a centralized management style

Source: *Scott-Berry, 1996a; Ryburg, 1996a; Keppler, 1997*

a minimal set (two to three) of standard workspaces with mobile or systems furniture, power walls or standardized components (Considine, 1999). These should support the various job functions while allowing for many options for future reconfigurations. The facility manager should aim to design the workspace with the fewest amount of modular components while providing the maximum amount of work settings.

4.4 Anticipating change

An absolutely essential element in space planning is anticipating the need to accommodate future technology changes in the company. Technological changes in the office infrastructure, combined with space reconfiguration, can enhance the effectiveness of a work environment. When facilities are built or upgraded, companies should attempt to forecast their information technology needs for the next five to ten years to make sure the new or upgraded space can meet those needs. Predictions beyond this are made impossible with the pace of technological change (Sullivan, 1995).

The power distribution system is an important part of designing and managing office facilities. Cabling infrastructure connects and must sufficiently power existing and future workstation phones, personal computers, local area networks, modems and fax machines, and should meet demands for speed, reliability and capacity (Scott-Berry, 1996a). A way of allowing for flexibility in an office is standardizing such infrastructure elements as cabling, lighting and ventilation systems within a module. Within a universal office plan, fixed infrastructure elements and movable wall systems can be transformed into various configurations. However, standardization should not get in the way of accommodating people. Rather than just dividing up space, determine what the work requires (Sullivan, 1995).

The use of flexible wall systems is a way of coping with frequent corporate changes as they are flexible, affordable and high performing. They are also environmentally responsible alternatives to fixed walls as they are generally reusable. They are also considered in the furniture and fixtures category, so they can be depreciated. Pre-finished panels allow faster completion and earlier occupancy. Using movable partitions or walls means workspaces no longer have to be planned out months or years in advance of their expected need. To determine the right movable wall partition for their needs, facility managers should use their office churn rate as a guide – the greater the change in office needs, the greater the degree of mobility should be incorporated in the office. Movable wall systems are most suitable if the annual churn rate is 20% or more (Tatum, 1996).

The universal plan strategy which originally involved moving workers from one workstation to another identical one has evolved into a more flexible system that allows for differences in work setting needs. The flexible universal plan has resulted from innovations by furniture manufacturers giving the user more control. Furniture components mounted on casters give complete mobility whereby panels or screens can be moved anywhere along a spine of panels that resembles a wall (Considine, 1999). If the spine is equipped with power and telecommunications technology, it serves as the core utility wall of the system. The user can create any configuration as required, without the need for a furniture consultant or facility manager to rearrange the system. This can be done for a short-term arrangement, such as a meeting, or a private space for concentration time. Within an hour these work areas can be created and can work for any length of time depending on the need. The ability to move the furniture as well as the occupants creates a truly flexible office environment.

To accommodate technology changes and restructuring, office furniture must be adjustable so that it does not become obsolete with a change in equipment. It may be good practice to limit the furniture palette to the fewest number of pieces as possible, and to consider furniture mounted on casters for maximum mobility and flexibility (Vergetis-Lundin, 1998). For tight office areas, a vertical space configuration (i.e. shelves) may be utilized for placing and storing technology components and current project documents (Ryburg, 1996b). This increases the available work surface area.

4.5 Space audits

Organizations are constantly changing, both from external influences over which they have little control and internal pressures, to become more efficient and competitive. A good example is the impact of office automation. The effect of automation is pervasive and influences:

1. The ergonomics of the workplace.
2. The amount of work surface required for personal screens, as well as the additional floor space required to accommodate printers and central processing units.
3. The levels of lighting and the design of light fittings to reduce glare and reflection on screens.
4. The capacity of the building and furniture to cope with wires and cables.
5. The need to extract heat from equipment.
6. The need for acoustic barriers to reduce the clatter of machines.
7. The pattern of work, and eventually the distribution of space.
8. The desire for a visual language that can cope with the proliferation of equipment and cables.
9. The cost of operating the building.
10. The cost of a new building, refurbishment and fitting out (Bernard Williams Associates, 1999).

The space audit provides an opportunity to review facility policies, take stock of the changing needs of the organization, test current space usage and assess the effectiveness of management. The most useful evaluation occurs when standards or benchmarks exist against which performance can be measured. It should be undertaken in consultation with specialist planners and designers.

4.6 Important considerations

In order for organizations to achieve a cultural transformation, a complete transformation of the physical workplace is required. Some important and generic guidelines comprise:

1. Eliminate the over-sized, status-based office.
2. Reallocate remaining space as formal and informal meeting places.
3. Provide adequate space for displayed thinking, ideas and communicating.
4. Equip a 'war room' with state-of-the-art technology.
5. Maintain consistency in furnishings and equipment company-wide.
6. Create excellent quality food and drink team areas; eliminate individual service.
7. Encourage and enable team collaboration and communication.
8. Provide resources to support explicit and conscious change management (Sims, 1998).

Perhaps the key strategies in effective space planning are best summarized by Trickett (1998). He states eight strategies that need to be pursued:

1. Measure needs by evaluating people's wants.
2. Reflect management style by studying the ways people work together.
3. Recognize social needs by removing barriers to communication.
4. Encourage interaction by helping people to work in groups.
5. Improve efficiency and comfort by re-engineering workplace tools.
6. Influence people's attitudes by defining the surroundings in which they work.
7. Increase variety by expressing people's differences.
8. Contribute to people's 'self esteem' by focusing on their sense of well-being.

Space utilization and best facility management practices at the planning level require consideration of further issues, such as:

1. Shifts from working alone to more collaborative teamwork are generating pressure for more available break-out meeting spaces and access to technology in meetings.
2. The increasing need to efficiently respond to changes in floor plan layout, makes it more important to incorporate flexible partitioning systems and carefully weigh primary and secondary routes for cabling.
3. The allowance for technology support and distribution, such as telephone, data and electrical closets, is more complex and frequently larger because of cable, equipment and service access needs.
4. The need for more and larger shared use technology spaces for larger laser printers, copiers, CAD stations, fax machines and other equipment (Ryburg, 1996b).

The challenge for facility managers is to provide spaces that are flexible and adaptable to change. Issues of noise reduction and visual privacy remain difficult to solve in open plan work environments, and will be made more significant with the introduction of voice-activated computer technologies in the years ahead.

References and bibliography

Alexander, K. (1996). *Facilities Management: Theory and Practice*. E. & F.N. Spon.
Atkin, B. and Brooks, A. (2000). *Total Facilities Management*. Blackwell Science.

Barrett, P. (1995). *Facilities Management: Towards Best Practice*. Blackwell Science.
Becker, F. and Steele, F. I. (1995). *Workplace by Design*. Jossey-Bass.
Bernard Williams Associates (1999). *Facilities Economics*. Building Economics Bureau Limited.
Considine, M. P. (1999). 'Innovations Boost Flexibility: New products are giving users more control of office space'. *Building Operating Management*, May.
<http://www.facilitiesnet.com/fn/NS/NS3b9ej.html>
Cotts, D. G. (1999). *The Facility Management Handbook* (2nd edition). AMACOM.
Damiani, A. S. (1998). *Moving up the Organization in Facilities Management: Proven Strategies to Increase Productivity in your Workforce*. Scitech Publishing.
Green, G. M. and Baker, F. (1991). *Work, Health, and Productivity*. Oxford University Press.
Guzzo, R. A. and Dickson, M. W. (1996). 'Teams in Organizations: Recent research on performance and effectiveness'. *Annual Review of Psychology*, 47, pp. 307-338.
Hamer, J. M. (1988). *Facility Management Systems*. Van Nostrand Reinhold.
Henderson-Paradis, R. (1996). 'The Strategic Planning Process and the CEO: Facilities managers need the tools, technology and new processes to provide top management with important information'. *AFE Facilities Engineering Journal*.
<http://www.facilitiesnet.com/fn/NS/NS3a96e.html>
Hesselbein, F., Goldsmith, M. and Beckhard, R. (1997). *The Organization of the Future*. Jossey-Bass.
Keppler, K. (1997). 'Bridging the Gap between Work and Workplace: Careful selection of an alternative workspace strategy is a key part of the process of transforming an organization'. *Building Operating Management*.
<http://www.facilitiesnet.com/fn/NS/NS3b7ha.html>
McGregor, W. and Then, D. (1999). *Facilities Management and the Business of Space*. Arnold.
Moos, R. H. (1996). 'Understanding Environments: The key to improving social processes and program outcomes'. *American Journal of Community Psychology*, 24, pp. 193-201.
Pacanowsky, M. (1995). 'Team Tools for Wicked Problems'. *Organizational Dynamics*, 23, pp. 36-51.
Park, A. (1994). *Facilities Management: An Explanation*. Macmillan.
Price, S. (1997). 'Facilities Planning: A perspective for the Information Age'. *IIE Solutions*, August, pp. 20-22.
Rondeau, E. P., Brown, R. K. and Lapides, P. D. (1995). *Facility Management*. John Wiley & Sons.
Rosenblatt, B. (1995). *New Changes in the Office Work Environment: Toward Integrating Architecture, OD, and Information Systems Paradigms*. Ablex Publishing Corporation.
Ryburg, J. (1996a). 'Alternative Office Strategies: New work patterns are shaping office design'. *Building Operating Management*, March.
<http://www.facilitiesnet.com/fn/NS/NS3b36a.html>
Ryburg, J. (1996b). 'Office Automation: Space planning practice: Best practice findings'. *FacilitiesNet*.
<http://www.facilitiesnet.com/fn/NS/NS3pdtm5.html>
Salvendy, G. (1997). *Handbook of Human Factors and Ergonomics* (2nd edition). John Wiley & Sons.
Schulze, P. C. (1999). *Measures of Environmental Performance and Ecosystem Condition*. National Academy Press, Washington.
Scott-Berry, E. (1996a). 'Workplace Planning Strategies: Flexibility – new work styles need adaptable workspaces'. *Building Operating Management*, August.
<http://www.facilitiesnet.com/fn/NS/NS3b86e.html>
Scott-Berry, E. (1996b). 'Workplace Planning Strategies: Identity – Image goes beyond the physical building'. *Building Operations Management*, August.
<http://www.facilitiesnet.com/fn/NS/NS2bmaug.html>
Sims, W. R. (1998). 'Death of the Executive Row'. *Facilities Design and Management* 17(11), November, p. 44.
Stevens, D. (1997). *Strategic Thinking: Success Secrets of Big Business Projects*. McGraw-Hill.
Sullivan, E. (1995). 'Does Office Design Hinder Employee Productivity? The answer is yes, according to many corporations. Here's what they're doing about it'. Trade Press Publishing Corporation.
<http://www.facilitiesnet.com/fn/NS/NS3bh5a.html>
Tatum, R. (1996). 'Movable Objects: Flexible wall systems help meet demands of a changing workplace'. *Building Operating Management*, August.
Tatum, R. (1997). 'Tackling Workplace Problems: The physical environment can boost or limit productivity in seven key areas', *Building Operating Management*.
<http://www.facilitiesnet.com/fn/NS/NS3b7ec.html>

Tompkins, J. A. (1996). *Facilities Planning* (2nd edition). John Wiley & Sons.
Trickett, T. (1998). 'Transforming Organizational Life by Design'. In *Facilities Management: Theory and Practice* (K. Alexander, ed.), E. & F.N. Spon, pp. 14-27.
van Delinder, T. (1997). 'Creating a Progressive Office Environment'. *Facilities Management Journal*, May/June, pp. 18-22.
Vergetis-Lundin, B. L. (1998). 'Designing Productive Office Space: Keep in mind that corporate image, technology and the need for team spaces play a role when building high-performance workspaces'. *Building Operating Management*, August. <http://www.facilitiesnet.com/fn/NS/NS3b8ha.html>

Needs analysis

5.1 Introduction

Asset characteristics and corporate policies regarding service levels must be thoroughly understood and documented in order for the facility manager to develop fixed operating plans, schedules and resource requirements with a high degree of accuracy. Operations that are impacted primarily by changes in product and business plans can be anticipated and allowed for by observing key business indicators, staying abreast of the dynamics of the organization and participating in decision-making.

The impact of change varies with the scope of the operation and the magnitude of the change. Organization size and complexity, the size and distribution of the employee population, and the quantity, size and distribution of facilities are key factors. Any domino effect (i.e., one change leading to another change) must also be properly considered (Fuller, 1998). Technological developments are a good example of rapid change having a widespread impact on the organization.

Information needed to anticipate facility requirements is regularly gathered and analysed by developers of business strategy, product planning and competitive (marketing) analysis. Communication links with finance, personnel and organizational planning groups allow facility requirements to be identified and alternative strategies to be discussed as these plans are developed. An appropriate proactive and flexible facility plan results from a joint effort of all affected parts of the organization. It is critical that developments in infrastructure, while reflecting and supporting overall strategic directions, also take account of organizational needs.

5.2 What is a needs analysis?

Before implementing any new facility initiative, it is essential that a plan is formulated based on a business needs analysis. For the plan to be successful and strategic, it has to be aligned with overall organizational needs, as well as have management involvement. It is also important that benchmarking is used to compare performance with similar organizations or industry averages, although this incurs a lag time in relation to the benchmarked organization.

A needs analysis is really two processes: integrating organizational plans and auditing performance. Long-term plans result from the setting of goals and objectives. These plans have implications for facilities, but increasing technology is providing new ways to do things that can improve the likelihood of success of these goals and objectives. Integration of business goals and infrastructure development is therefore critical in today's society.

The audit process is divided into four areas: technology, work flow, paper flow and resources. These audits provide an analysis of existing conditions and help determine future requirements for functionality and procedures. Audits also determine if some current procedures of an organization, especially those based on resources, are inefficient. The audit process lets the company re-evaluate work procedures based on tasks, not resources.

5.2.1 Information gathering

Needs are determined from an analysis of strategic objective implementation, but in effect their assessment is linked to the expectations of customers. This information forms the basis to identify and analyse all practical facility alternatives that the market can provide. During this important first phase the following should be determined:

1. Locational preferences.
2. Timing constraints.
3. Space required at occupancy.
4. Growth/contraction projections.
5. Image desired.
6. Occupancy cost objectives.
7. Performance criteria for building systems.
8. Space utilization/efficiency/adjacencies.
9. Optimum floor configuration and size.
10. Amenities required/desired. (Brown et al., 1994)

5.2.2 Analysis phase

In the analysis phase, the facility manager should evaluate the customer needs data and identify all viable alternatives that are available in the market. Where acquisition of new facilities is necessary, each option should be analysed according to:

1. Locational considerations and access.
2. Proximity to labour/clients/competitors.
3. Operational considerations.
4. Time required for implementation.
5. Total net present value occupancy costs.

6. Detailed construction specifications and costs.
7. Architectural efficiency of alternative buildings.
8. Comprehensive evaluation of building systems.
9. Site/parking/amenity evaluation.

The analysis should be summarized to provide a concise and objective presentation of relevant considerations attending each option. The customer can then proceed more rapidly through the decision-making process and objectively weigh the economic and functional strengths and weaknesses of each alternative.

5.2.3 Strategic development

The importance of strategic development cannot be underestimated in the process of negotiating a real estate transaction. To achieve the best terms that dynamic market conditions will allow requires:

1. Developing maximum negotiating leverage.
2. Understanding each landlord's unique agenda.
3. Recognizing changing market conditions.
4. Defining risks.
5. Controlling privileged information.

Managing change is the single largest issue facing senior management today. The objective here is to help the customer's organization make real estate decisions which accommodate potential future changes such as expanding staff, restructuring, or contracting with minimum penalties. Other information needed comprises:

1. Marketing distribution plans, inventory management and sales forecasts by product/service type.
2. Financial objectives including funding, profitability and profile.
3. Organizational structure for operating, regulatory and liability management.
4. Human resources, as related to site-specific operating costs and relocations.
5. Real estate inventory and utilization data, historical and forecasted.
6. Strategic and operating requirements for each strategic business unit.
7. Ancillary services related to facility operations (utilities, amenities, etc.).
8. Market information respecting the commodity (space and location).
9. Real property interest preferences (fee or leaseholds).
10. Financing methods: internal, seller financing, third party (for example, banks), or leases.
11. Financial statement objectives: asset, liability or expense recognition and timing.
12. Cash flow impact.
13. Management of financial, market and legal risks and contingencies.
14. Strategic aspects of the transaction (for example, image, employee morale, environmental protection).

5.2.4 Implementation

Project implementation requires a team of professionals skilled in negotiation tactics, space programming, interior design, construction management, contract law and facility operations. When the corporate facility manager has achieved maximum landlord concessions and tenant contractual rights, the project can then be managed, with the probable assistance of external consultants, to protect those gains during the important phases of design, construction and relocation. The following checklist applies:

1. Finalize lease negotiations.
2. Prepare final space plans.
3. Engage consultants.
4. Prepare designs.
5. Supervise construction document preparation.
6. Negotiate construction pricing and contracts.
7. Administer construction to control costs.
8. Negotiate contracts for physical move.
9. Schedule the relocation.

Implementation is often outsourced under the control of the facility manager, who acts as the client representative. Some larger organizations have their own project management staff on board; in other cases, a consultant is engaged.

5.3 Strengths, weaknesses, opportunities and threats

Strategy is the basis for all sales and marketing planning, for new product or service development, for personnel development, and for the allocation of resources. Developing an effective strategy is the key to profitable survival and growth in an uncertain business climate. A useful technique that will help determine the actions required for a business to achieve its objectives and clarify viability and direction is called a SWOT analysis. The term SWOT is an anagram that stands for Strengths, Weaknesses, Opportunities and Threats.

A SWOT analysis evaluates the strengths and weaknesses of internal forces required for implementation and the external opportunities and threats offered by further development. A SWOT analysis is a corporate planning technique that enables future directions to be evaluated in the light of past successes and failures and new gains and losses. A list is prepared for each heading based on the collective input of the planning group.

Strengths are what the organization does well, or more particularly what the organization does better than most of its competitors. Examples of strengths may include image, turnover, employee satisfaction, facilities, proximity to clients, etc. Strengths are identified so that new ventures can capitalize upon them and use them to further increase market position.

Weaknesses are areas of business activity that are poor and require improvement. Examples of weaknesses may include lack of computer resources, budget constraints, no offshore links, high turnover of key staff, etc. Weaknesses can undermine the successful aspects of business activity and can potentially lead to takeover or restructuring. Weaknesses must be recognized and redressed.

Opportunities are new initiatives that if pursued would lead to improvements in financial position, market share and future proofing of the organization. Examples of opportunities may include entering new markets, launching a new product range, operating internationally, implementing e-mail throughout the organization, etc. Each opportunity needs to be further evaluated.

Threats are generally external effects over which the organization has little control. Examples of threats may include lowering of tariffs, increased interest rates on loans, takeover bids, exposure to litigation, etc. Threats cannot normally be overcome, but strategies may be able to be developed which limit their impact or plan for their management. Threats may lead to aggressive market activity.

A SWOT analysis is a group activity. It is normally undertaken in a 'brainstorming' environment. A SWOT workshop requires a leader or facilitator and ideally representatives from all key parts of the organization. External members may also be appropriate. All members of the workshop should have equal rights to contribute.

An internal analysis of strengths and weaknesses could include:

1. Services and products.
2. Market share and expansion.
3. Profitability.
4. Financial resources and level of borrowed funds.
5. Staff and management knowledge, skills and experience.
6. Computer capability and literacy.
7. Ability to prepare promotional material.
8. Ability to plan and execute effective business development programmes. (McLure, 1993; Roth and Washburn, 1999)

External research of opportunities and threats could include:

1. Major competitors – their present and future share of the market, and their strengths and weaknesses.
2. Politics – tax, environment, wages and regulations.
3. Economics – the state of the economy, interest rates and level of employment.
4. Social situation – changes in social behaviour, attitudes, standard of living and levels of incomes.
5. Technology – changes to existing products, methods of distribution and production.
6. Health and safety – changes in zoning regulations and/or pollution standards. (McLure, 1993; Roth and Washburn, 1999)

Following the analysis, the manager should be able to determine the organization's position and decide which steps to take and which resources have to be acquired in order to achieve business objectives. The strategies that are to be employed to maximize strengths and take advantage of competitors' weaknesses should be ascertained. A strategy developed in this way will promote response to competitive activity in an effective and proactive way.

5.4 Managing change

In a business climate of constant changes due to advances in technology, global competition, mergers, takeovers, strategic alliances, office churn and other business restructuring, striving to achieve or maintain a competitive advantage is a challenge. Organizations strive to deal with instability due to these changes through attempts to improve productivity, efficiency, and cost containment which require flexibility and innovation in the workplace. An unsuccessful approach would be to respond reactively to new demands and implcment a plan with no clear mission. Instead, a strategy in keeping with the organization mission needs to be devised so that the facility can respond proactively to change with a flexible infrastructure. This strategy must be reviewed when business conditions or objectives change (Carpio, 1997).

Facility managers need to support the organization's needs by facilitating staff needs for appropriate workspace and access to technology. They have to develop the systems, procedures and processes required to remain responsive to change while managing costs, time and effectiveness (Bell and Keppler, 1997). The ever-present emphasis on 'space' performance must not come at the expense of 'human' performance as there is much evidence that providing adequate facilities for the needs of workers almost always pays for itself in terms of productivity (Brand and Syfert, 1998). This requires specific strategies to balance competitive requirements of getting more people into less space, speed, flexibility, cost and the human element. Fast changes mean that workers can begin being productive immediately following any reconfiguration. Flexibility not only influences the speed of changes, but can also eliminate the need for radical re-planning.

Facility managers can accommodate high rates of change by implementing an infrastructure that allows for quick formation and dissolution of functional work groups and teams, and speedy data collection and access to information. Rather than spend time reconciling long-term corporate planning with long-range space and technology requirements (which will need numerous revisions), facility managers should concentrate on developing different alternative configurations that can be implemented quickly and at the lowest cost. According to Brand and Syfert (1998), there are four ways that facility managers can deal with change and not compromise employee performance and productivity:

1. Exploit to the fullest extent the space and other resources already available. This involves augmenting the functionality of a space without replacing the entire system (an interior designer may help to co-ordinate added components).
2. Design spaces rather than individual footprints. In order to accommodate more rapid change, work areas need to be designed to support the formation, function, attrition and re-combination of functional work groups. Facility constraints imposed on spaces should be flexible enough to allow the space to grow and adapt to changing needs.
3. Allow more user-centred control over the space available. Decisions about where to situate desks, tables, partitions, marker boards, chairs, telephones and computers can be given to individual workers. Local Area Networks (LANs) make individual locations interchangeable, and modular office furniture systems can be rearranged to accommodate changing needs.

4. Support both teams and private spaces within the same area. Design larger spaces that incorporate a variety of group levels and individual needs. Such work areas can be quickly reconfigured to co-ordinate and facilitate both teaming and private work.

5.5 Churn

Churn is defined as the turnover/movement of people and their office space. Churn rate refers to the number of people and their office spaces moved or changed within a period of time (usually one year) divided by the total number of staff. Churn occurs in many different ways. A churn rate of 100%, for example, could mean all of the people and all of the space, or it could mean the equivalent of all the people and space, i.e., 20% of the people and spaces moved five times in one year. A space move/change could mean moving the entire space, or it could mean the number of square metres on the floor that were affected by any kind of move/change to the furniture inside an office space. A churn rate can represent an aggregate of many different kinds of churn (Ryburg, 1996a).

Three primary sources of churn are:

1. Company-wide restructuring due to mergers and downsizing.
2. Re-engineering that causes companies to move groups of employees that work together near each other.
3. Ongoing formation and operation of project teams across the company (Ryburg, 1997).

Although companies need to embrace churn if they want to achieve their new goals, the facility side of churn can be unnecessarily disruptive and costly if appropriate strategies are not implemented. Three of the facility manager's biggest challenges in handling churn are related to walls (space), technology (particularly cables) and furnishings. While furnishings are moved the most frequently, and permanent walls the least often, the order is reversed when those three categories are ranked in terms of greatest cost and difficulty.

To reduce churn-related facility costs (and downtime disruptions to company operations and departments), facility managers use the following:

1. Forecast long-range and near-term requirements. Objective: reduce future churn. Method: strategic facility planning. This is the most effective way to reduce future churn rates and related costs following organization-wide restructuring.
2. Move people, not facilities. Objective: eliminate current churn. Method: free address or universal planning/templates in which employees can create any work configuration they need (walls, furniture and technology are all more adjustable).
3. Move everything except technology. Objective: reduce technology churn. Method: use spine-walls or raised floors/accessible plenums to provide for the primary distribution of cables.
4. Move everything including technology. Objective: embrace all churn. Method:

use raised floors/stand-up plenum spaces; large, open and unencumbered floor; and dedicated or roving (on-call) facility management furniture.

5. Adapt instead of move. Objective: keep churn localized. Method: flexible de-mountable walls/modular partitions, variable density layouts, variable screening, flexible work support, adjustable adjacency, universal templates to circumvent the need to change the furniture and panels altogether using off-module furniture products. Furniture systems must be able to accommodate new conditions and adapt to new requirements.
6. Electronic, not physical, moves. Objective: virtual vs. facility churn. Method: desktop tele-teams, work from anywhere. The use of desktop and other teleconferencing technologies to facilitate team collaborations is increasingly viewed as a viable alternative to the physical co-location of workers in one spot. Connecting people electronically at the desktop rather than physically not only reduces people and facility churn at team sites, but the cost of maintaining electronically co-located teams is far less expensive than physical solution strategies (Ryburg, 1996b; Ryburg, 1997).

Organization-wide restructuring always involves the largest number of people, and the most effective facility management control for this level of churn is accurate forecasting, particularly of changing facility capacity and location requirements. Fifty per cent of all churn which occurs within two years of major restructuring can often be traced to inadequate facility (capacity and location) implementation at the time of restructuring. Churn caused by ongoing co-location shifts is generated primarily by an organization-wide drive to achieve (1) greater efficiencies within and between operations/departments, and (2) a team-based organization. The most effective facility management control is to ‘move people not facilities’, focusing on ways to keep the interior space envelope more flexible by simplifying wire/cable distribution, the use of flexible de-mountable furniture, etc. Churn due to ongoing teaming operations is the result of many team formations, starts, stops and changes that occur in teams over the course of a project. The most effective facility management controls are those that respond easiest to shifts in the layout of team spaces and home base offices, with a minimum of wall and utilities (wire/cable) change, as in universal planning with off-module capabilities and more adaptive/moveable furniture elements (Ryburg, 1996a).

Some companies sustain high rates of churn because they feel it engenders continuous improvement. The financial gain is worth bearing the cost of churn, but companies still seek to minimize costs by making facilities flexible.

References and bibliography

Alexander, K. (1996). *Facilities Management: Theory and Practice*. E. & F.N. Spon.

Atkin, B. and Brooks, A. (2000). *Total Facilities Management*. Blackwell Science.

Barrett, P. (1995). *Facilities Management: Towards Best Practice*. Blackwell Science.

Bell, S. and Keppler, K. (1997). ‘Strategy and Integration: The FM’s broader charge’. *Facilities Design and Management*, 16(10), October, pp. 59-61.

Brand, J. L. and Syfert, T. (1998). ‘Facilities Strategies to Support Corporate Change and Flexibility’. *Office Trends*, February. <http://www.haworth-furn.com/resource/ot/factat.html>

Brown, R. K., Lapides, P. D. and Rondeau, E. P. (1994). 'Corporate Policy is Part of RE/FM Planning'. *Facilities Design and Management*, 13(7), July, pp. 50-53.

Carpio, C. E. (1997). 'Thriving in Uncertain Times: Flexibility and innovation key'. *AFE Facilities Engineering Journal*, July/August.

Cotts, D. G. (1999). *The Facility Management Handbook* (2nd edition). AMACOM.

Damiani, A. S. (1998). *Moving up the Organization in Facilities Management: Proven Strategies to Increase Productivity in your Workforce*. Scitech Publishing.

Fuller, M. J. (1998). 'Attack your Strategic Plan in Two Prongs'. *Facilities Design and Management*, 17(3), March, pp. 54-57.

Green, G. M. and Baker, F. (1991). *Work, Health, and Productivity*. Oxford University Press.

Guzzo, R. A. and Dickson, M. W. (1996). 'Teams in Organizations: Recent research on performance and effectiveness'. *Annual Review of Psychology*, 47, pp. 307-338.

Hesselbein, F., Goldsmith, M. and Beckhard, R. (1997). *The Organization of the Future*. Jossey-Bass.

McGregor, W. and Then, D. (1999). *Facilities Management and the Business of Space*. Arnold.

McLure, B. (1993). *The Small Business Handbook: How to Start and Successfully Operate a Small Business*. Information Australia, Melbourne.

Moos, R. H. (1996). 'Understanding Environments: The key to improving social processes and program outcomes'. *American Journal of Community Psychology*, 24, pp. 193-201.

Park, A. (1994). *Facilities Management: An Explanation*. Macmillan.

Rondeau, E. P., Brown, R. K. and Lapides, P. D. (1995). *Facility Management*. John Wiley & Sons.

Rosenblatt, B. (1995). *New Changes in the Office Work Environment: Toward Integrating Architecture, OD, and Information Systems Paradigms*. Ablex Publishing Corporation.

Roth, B. N. and Washburn, S. A. (1999). 'Developing Strategy'. *Journal of Management Consulting*, 10(3), May, pp. 50-54.

Ryburg, J. (1996a). 'Understanding the Dynamics of Churn: Change produces different kinds of churn in different kinds of organizations'. *FacilitiesNet.* <http://www.facilitiesnet.com/fn/NS/NS3chrn1.html>

Ryburg, J. (1996b). 'The Best Facility Organizations have Found Ways to Limit the Costs and Disruption of Churn'. *FacilitiesNet.* <http://www.facilitiesnet.com.fn/NS/NS3chm2.html>

Ryburg, J. (1997). 'Managing the Impact of Churn'. *Building Operating Management*, January. <http://www.facilitiesnet.com/fn/NS/NS3b7aa.html>

Salvendy, G. (1997). *Handbook of Human Factors and Ergonomics* (2nd edition). John Wiley & Sons.

Schulze, P. C. (1999). *Measures of Environmental Performance and Ecosystem Condition*. National Academy Press, Washington.

Stevens, D. (1997). *Strategic Thinking: Success Secrets of Big Business Projects*. McGraw-Hill.

Tompkins, J. A. (1996). *Facilities Planning* (2nd edition). John Wiley & Sons.

van Delinder, T. (1997). 'Creating a Progressive Office Environment'. *Facilities Management Journal*, May/June, pp. 18-22.

Portfolio management

6.1 Introduction

The focus for facility managers has moved away from managing the budget to being accountable for managing an asset and increasing its value – from the tactical to the strategic (Kroll, 1997). In order to become a good investment manager over the long term, a strategic approach must be adopted to repeatedly deliver good portfolio returns. A systematic strategy should give facility managers the ability to refine their judgement, dissect positive and negative outcomes, and learn to adjust their strategy from past performance. The strategic approach considers all the markets and decides which markets to consider and how much time is to be spent in those markets to increase the probability of obtaining the best deal overall (Viezer, 1999).

According to Viezer (1999), the typical annual cycle for a facility manager would include the following tasks:

1. The previous year's portfolio performance is reviewed and analysed.
2. The market outlook for the next year is forecast, and then markets targeted for investment acquisition, hold and disposition.
3. A range is developed providing the optimal asset allocations. Groups of assets are weighted to produce a portfolio that maximizes the expected return for a given level of risk.
4. Recommendations are made for individual properties on how to 'beat' the forecasted market returns.
5. The markets and portfolio performance are continuously monitored. The outlook and target markets are updated and revised if necessary before developing the next annual plan and budget.

Targeting markets for investment using a disciplined and systematic approach can have several benefits. It can provide more control to facility managers, allowing them to efficiently allocate the acquisition of staff resources, provide an opportunity to improve performance, and complement portfolio diversification.

6.2 Portfolio diversity

Portfolio theory decrees that investors should select the best combination of investment media to either maximize return for a given level of risk or minimize risk for a given level of return (Rubens et al., 1998). The primary motive in holding a diversified portfolio of securities is to reduce risk. Investors holding a diversified portfolio expect to reduce the risk associated with earning a return that is less than the available return (Geurts and Nolan, 1997). The total risk of the portfolio will depend not only on the number of securities in the portfolio, but also on the risk of each individual security and the degree to which these risks are independent of each other. Benefits of a portfolio will be improved if diversification is based on a combination of property types, cities in which property is located, growth rates and lease maturity (Wolverton et al., 1998; Geurts and Nolan, 1997).

Fortunately, there are portfolio diversification tools that help facility managers determine which markets offer the best prospects for investment (Viezer, 1999). They assess markets and the amounts that should be included in a portfolio to maximize the expected return for a given level of risk. Some tools calculate a number of portfolios that maximize expected return over a range of risks. These tools can calculate the rate of return that additional investment must achieve in order to increase the risk-adjusted return of the portfolio. They can also be used to evaluate individual deals and to suggest strategy, with uncertainty incorporated into the analyses to provide flexibility to decision-makers.

6.3 Managing company assets

Outsourcing and improved operational management used to free capital tied up in facilities, improve quality and reduce costs, is being replaced by new, innovative facilities and property solutions which require the facility manager to understand the financial potential in facilities. A chasm is growing between business needs and property offerings as less fixed space is being required by major organizations. The space required is becoming more flexible, and of a much higher specification. Consequently, when servicing the property portfolio of clients or employers by optimization of space within it, instead of the traditional method of minimizing space, the facility manager may consider an alternative option of releasing capital. Exchanging freehold and leasehold interests for more flexible service occupancy agreements in the same locations but with value being released in the form of cash, reduced running costs and real future space flexibility, frees them of the volatility and constraints associated with direct property interests (Jones, 1998).

At the February 1997 British Property Federation national conference, it was suggested that the full repairing and insuring lease will eventually be replaced by 'service occupancy contracts, with both the risks and rewards of ownership transferred to the owner' with securitization used as a way to fund projects. Securitization creates a marketable interest in a property or group of properties. Investors acquiring a stake will hold equity or a security in the product, which is of course a stake in the income or rent generated. This is a clear concept but not one that immediately yields great step changes for the facility manager or the occupier, because investors in a traditional property company are only interested in the reliability and growth potential in the rental income stream. This is assessed against other potential investment solutions – gilts, money markets or general shareholdings (Jones, 1998).

Real estate and personal property (such as fixtures, furnishings and equipment) are significant assets in most businesses, but their true value and availability are often overlooked. Thorough understanding and documentation of asset characteristics and uses and corporate service requirements are needed for the facility manager to develop sound operating plans, schedules and resource requirements. All stages of the life cycle (i.e., procurement, assignment, delivery, use, reallocation, and eventual disposition) must be considered in the facility strategic plan. Otherwise, without co-ordination of data, new equipment could be purchased while equivalent items are being released from an area or sitting idle in a warehouse. As another example, a department may obtain leased space for expansion without realizing that an adjacent department is planning to vacate. The alert facility manager can intervene and develop alternatives to meet requirements with minimal cost and disruption (Fuller, 1998).

There has been a gradual trend in the real estate portfolios of organizations from cost-centred commodities to strategic assets that can be dynamically allocated and managed to support organizational objectives. Information technology and a sophisticated communication environment have influenced organizational processes and individual work patterns, and consequently the workplace is being seen as an important strategic tool in achieving the organization's performance goals. Hence, it is in the interest of large building stock owners to evaluate the suitability of their property holdings so as to meet their organizational and technological needs and their environmental and occupant demands (Mahdavi and Shankavaram, 1995).

During economic recessions many organizations put off scheduled maintenance that eventually has to be dealt with. Software products exist that take the building deficiency data collected during a physical audit and combine it with other data such as cost spreadsheets, financial investment models and computer-aided facility management (CAFM) software. Such integrated software, which can incorporate multimedia information, is a powerful forecasting and strategic planning tool for management, and can be used for real property and facility assessment purposes.

Condition assessment is based on a physical building audit in which deficiencies are identified, and how the deficiencies are to be corrected is determined along with the cost of those repairs (Teicholz, 1995). Condition assessment data, traditionally resulting in a written report, is now being used for ongoing management with software systems developed to support both data collection and flexible reporting. It is possible to acquire condition assessment software that provides the user with a mechanism for entering updated information as deficiencies are corrected, resulting in an ongoing management tool for planning purposes.

The linking of condition assessment systems with CAFM offers powerful tools for tracking and reporting on the results of building condition assessments. The benefits of condition assessment systems linked to CAFM systems are numerous and have both strategic and tactical implications for organizations. At a tactical level, the systems can determine, in a highly flexible manner, capital implications and costs of various types of deficiency associated with the physical plant. It can also be used for regulatory documentation and to justify indirect cost recovery requests. At a strategic level, the systems provide a dynamic database that can be used for long-range facility forecasting of deferred maintenance costs. Using the cost templates stored in the database, modified for specific conditions and using various depreciation schedules and replacement costs associated with various systems, an accurate database of information can be created and maintained. In this manner, management can model various investment strategies and predict costs for optimal investment in the physical plant or forecast the implications of not making an investment in plant. Management can also look at the potential liabilities of various organizational units based on their facilities and various types of asset associated with those units.

6.4 Tracking industry and technology trends

Facility managers face both tactical and strategic challenges. Some of the tactical issues with which they have to face are:

1. Getting out of leases.
2. Disposing of real estate.
3. Renegotiating rents.
4. Managing high rates of churn (office reconfiguration).
5. Reducing operations and maintenance expenses. (Hubbard, 1994)

In order to become more strategic, facility managers need to pay close attention to industry and technology trends and explore new concepts to accelerate the pace of change and the application of technology that will directly translate into advantages to the organization's core business. To be successful leaders in their companies, facility managers need to be able to carry out:

1. Facility strategic and tactical planning.
2. Financial forecasting and budgeting.
3. Real estate procurement, leasing and disposal.
4. Quality management, benchmarking and best practices.
5. Telecommunications and information technology, and more. (Tuveson, 1997)

Vigilant monitoring and response to changing technology and business trends is essential to assure property value and avoid obsolescence. In order to keep track and take advantage of industry and technology trends, the facility manager should ideally:

1. Be a change agent, become enthusiastic about technology, and learn to embrace change and be proactive. Be open to new ideas about the workplace in terms of technology and research information (for example, ergonomics, indoor air quality and lighting) to create highly productive workplaces.
2. Develop and maintain a people network, create strategic alliances, use experience-based outsourcing, be attentive to the profession and the direction of the organization, be willing to share information, and build a dependable network of world-class providers of products and services.

3. Be an entrepreneur, evaluate everything on performance, don't be trapped by your own experts, focus on 'what if?' instead of 'what I should have', and think creatively.
4. Be flexible. Facility managers must be flexible and responsive to market and technological trends. (Zulkeski, 1997)

The future will see more focus on increasing shareholder value, improving return-on-assets, and making the workplace an environment that meets all internal users' needs. Facility managers must also be able to support employees who work in alternative workplaces. To succeed in the future, facility managers should also:

1. Have an imaginative understanding of all of their internal customers' needs.
2. Develop effective information systems to improve access to information.
3. Foster a match between facility management and real estate strategies and their delivery processes. (Tuveson, 1997)

When the facility manager actively participates in the strategic planning process, decisions become integrated, proactive and effective. To do this properly facility managers should analyse market demands and industry trends. Sources of information can include independent and academic research groups, industry organizations and professional associations, consumer focus groups and vendors (Fuller, 1998).

As facility managers try to break away from a tactical role by outsourcing many daily operations and casting off assets, 'management attention must focus not only on cost but also on how to achieve sustainable competitive advantage to get the most of the strategic potential facilities can provide' (Hubbard, 1994). Jay McMahan of DEGW, London, suggests that in their strategic capacity, 'facilities managers must focus on how their efforts combine to (a) gain new customers, (b) retain existing customers, and (c) grow revenue' and 'carry out SWOT [strength, weakness, opportunity, threat] analysis in order to participate in real competitive strategy' (Hubbard, 1994) and to survive during sudden changes or turbulent times.

The facility manager has to deal with sweeping changes in the workplace as a result of a more informed and experienced workforce, new work methods and new technological tools. To enforce significant changes in place to follow these changes, facility managers should consider the following questions:

1. Can I readily absorb new electronic technology?
2. Are work processes continually breaking down?
3. Do changes in physical environments represent a crisis?
4. Does the environment support cost containment?
5. Have alternatives to traditional/territorial space been fully explored?
6. Am I meeting energy-efficiency objectives?
7. Does the operation of the workplace reflect corporate goals?
8. Are my environments ergonomically responsible? (Zulkeski, 1997)

References and bibliography

Alexander, K. (1996). *Facilities Management: Theory and Practice*. E. & F.N. Spon.
Atkin, B. and Brooks, A. (2000). *Total Facilities Management*. Blackwell Science.
Barrett, P. (1995). *Facilities Management: Towards Best Practice*. Blackwell Science.
Cotts, D. G. (1999). *The Facility Management Handbook* (2nd edition). AMACOM.
Fuller, M. J. (1998). 'Attack your Strategic Plan in Two Prongs'. *Facilities Design and Management*, 17(3), March, pp. 54-57.
Geurts, T. G. and Nolan, H. (1997). 'Does Real Estate have a Place in the Investment Portfolio of Tomorrow?'. *Review of Business*, 18(4), pp. 19-24.
Hamer, J. M. (1988). *Facility Management Systems*. Van Nostrand Reinhold.
Hubbard, G. M. (1994). 'Why you Should Become More Strategic and Less Tactical'. *Facilities Design and Management*, 13(2), February, pp. 50-53.
Jones, O. (1998). 'The New Property Paradigm'. *Supply Management*, 3(7), pp. 22-24.
Kroll, K. (1997). 'New Rules for Real Estate: As the game changes, so must the players'. *Building Operating Management*.
<http://www.facilitiesnet.com/fn/NS/NS3b7ib.html>
Mahdavi, A. and Shankavaram, J. (1995). 'A Progressive Resolution Method for the Evaluation of Large Building Stocks'. *Facilities*, 13(12), December, pp. 17-20.
McGregor, W. and Then, D. (1999). *Facilities Management and the Business of Space*. Arnold.
Park, A. (1994). *Facilities Management: An Explanation*. Macmillan.
Rondeau, E. P., Brown, R. K. and Lapides, P. D. (1995). *Facility Management*. John Wiley & Sons.
Rubens, J. H., Louton, D. A. and Yobaccio, E. J. (1998). 'Measuring the Significance of Diversification Gains'. *Journal of Real Estate Research*, 16(1), pp. 73-86.
Stevens, D. (1997). *Strategic Thinking: Success Secrets of Big Business Projects*. McGraw-Hill.
Teicholz, E. (1995). 'Computer-aided Facilities Management and Facility Conditions Assessment Software'. *Facilities*, 13(6), May, pp. 16-19.
Tompkins, J. A. (1996). *Facilities Planning* (2nd edition). John Wiley & Sons.
Tuveson, K. (1997). 'Keys for Success in the Year 2000 and Beyond'. *Managing Office Technology*, 42(12), December, pp. 22-23.
Viezer, T. W. (1999). 'Constructing Real Estate Investment Portfolios'. *Business Economics*, 34(4), October, pp. 51-58.
Wolverton, M. L., Cheng, P. and Hardin, W. G. III (1998). 'Real Estate Portfolio Risk Reduction through Intracity Diversification'. *Journal of Real Estate Portfolio Management*, 4(1), pp. 35-41.
Zulkeski, A. G. (1997). 'The FM as Change Agent'. *Facilities Design and Management*, 16(1), January, p. 64.

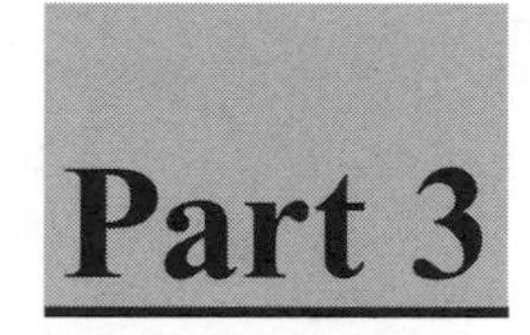

Information management

Knowledge is everything. Effective use of information can enable profit maximization and help increase quality, reduce cost and minimize risk. How an organization uses information and communicates it is critical to realizing these goals. Information may be qualitative or quantitative, but must be relevant, current and sufficiently detailed to support ongoing management decisions. Today all information management systems are computer-based, and increasingly web-enabled.

The key to successful information management is the underpinning technology and communication strategy. An open architecture will enable information to be used for a wide variety of applications and be available from different geographical locations. This permits sharing across particular facilities within an organization's portfolio and thereby permits management reporting to be centralized.

While the interpretation of data and resultant decisions are the ultimate outcome, poor base information will lessen the quality of these activities and in fact undermine the ability of the organization to learn from and react to changing environments, let alone be in a position to predict future events. Mechanisms need to be invented to capture current performance so that a complete picture is available at any point in time.

Facilities can be made more intelligent through the installation of computer-based management systems. These are not restricted to storage of information, but are able to monitor, collect, interpret and interact with physical systems to optimize performance. For example, systems which can shut down air conditioning and lighting zones when spaces are vacant will deliver higher operational savings and energy demand reduction, even though capital cost may be increased by the introduction of the necessary detectors and computer controls. Building management systems are common in medium to large-scale developments, but their potential is still relatively untapped.

Databases of information must be interconnected, giving seamless access to a wide cross-section of performance measures. There is much interest in the area of computer-aided facility management (CAFM) systems that link operational aspects via a three-dimensional building model. Information is entered and viewed using a graphical interface that represents the current design configuration and its contents. These are extremely powerful tools for the facility manager.

In order to enable CAFM systems to link to wider organizational databases, recent advancements in enterprise resource planning (ERP) technology are making it possible to integrate a diverse range of information so that managers are able to view overall performance and model new initiatives. While facility management is a part, the connection of traditionally autonomous groups offers substantial benefit. For example, the link between space management and human resource management enables innovative solutions such as hot-desking and hotelling to be successfully implemented.

As business becomes more competitive and more global, the role of information management will take on even higher priority. The days of paper-based systems are limited. The Internet has introduced new ways to communicate that are revolutionizing business. E-commerce is itself a recognized field and one that can enable organizations to expand their market and streamline operations. Many organizations have done this and have discovered the advantage which online information handling offers when linked to core business activities like sales, manufacturing, financial control, advertising, customer service and training. Information is indeed power.

This part looks at the role of information management in improving organization productivity. In particular, the impact of information management systems on the discipline of facility management is explored. Chapter 7 investigates communication strategies and the technology necessary to enable integration and data sharing. Chapter 8 focuses on building management systems and controls, including the notion of intelligent buildings that customize spatial environments as operational events dictate. Chapter 8 discusses the role of computer-aided design models as databases for facility information and as effective management tools.

Information management raises issues of access, security and privacy that must be satisfactorily addressed within an organization. However, appropriate design of solutions has the potential to transform the way in which an organization works, and lead to gains in productivity that will ultimately increase value to shareholders and/or other stakeholders, and improve customer service and satisfaction levels. Initial investment is necessary to realize future benefits, and this remains a major obstacle to their development.

Communication strategies

7.1 Introduction

Two of the most significant features that have arisen in and influenced the business environment have been the global nature of business activity and the exponential growth and convergence of computing, communications and media technologies. The ability to identify, locate and deliver information and knowledge to a point of valuable application has and will continue to transform businesses. As an increasingly mobile workforce, staggered work schedules and the need for instant communication transforms the office, the information co-ordinator has to manage vast amounts of information and the facility manager has to provide an infrastructure to support critical communication links among remote teams.

Organizations are realizing that to make the best decisions they need easy access to information across all levels of the organization. For instance, an organization with independent information technology systems for facilities and corporate management will find it difficult to achieve high operating efficiencies at both the facility management and corporate management levels. Corporate managers will not have ready access to facility information, such as maintenance, energy and operating costs, while facility managers will lack access to corporate information, such as up-to-date budget data, long-range plans for facilities, purchasing agreements, and staffing levels and costs. Organizations that have linked purchasing, inventory management and maintenance scheduling, for instance, are able to reduce the spare parts inventory by 35 to 50% (Piper, 1999b).

Advances in hardware and software are enabling facilities to link many functions to share data and communicate more effectively. Two related technology developments that allow linking of data systems are the Internet and the corporate intranet. These allow authorized users to access a wide range of data through a web browser and work with their own data. Internet-enabled computer-aided design (CAD) permits sharing software drawings with colleagues and clients over the Web.

Technology in the workplace has also become an integral part of effective communication between co-workers and clients. All of the developments in physical network infrastructures, speed and computing power are, in part, fuelled by changes in the way people work which in turn is made possible by the rapid expansion of the Internet, and use of corporation-wide and private e-mail. Technology has made it possible for people to set up offsite work environments, or virtual offices, anywhere they work. Telecommuters who work at home or at other remote locations, and roving workers who require mobile offices, are able to set up or carry everything, from fax and phone to computer, in order to work, as office equipment has become smaller and more easily transported. For the facility manager, the anytime-anywhere office means furniture and office equipment must be provided to employees according to their work patterns.

According to the International Facility Management Association, 'facility management is the practice of coordinating the physical workplace with the people and the work of the organization and integrates principles of business administration, architecture and the behavioural and engineering sciences' (Archibus, 1999). To cope with the tremendous amount of knowledge needed to solve a multitude of complex problems and challenges, and to deal with the co-ordination of all the details related to planning, designing and managing complex facilities, including systems, equipment, furniture and people, effective communication between the facility manager and other managers is critical.

To manage the increase in responsibilities for providing an effective and safe workplace for employees, facility managers should ensure they stay up to date with the latest technologies and strategies. Facility managers need software that help them to manage product and project information across the enterprise and optimize business results for greater profitability through more effective and efficient automated work processes. The procurement, operation and management of corporate facilities accounts for up to 50% of an organization's annual expenses and facility management automation can help realize up to 34% of costs (Archibus, 1999).

7.2 Infrastructure for successful communications and information handling

Computer and telecommunication systems have become critical elements in the day-to-day operation of businesses worldwide. As the technology that drives the systems has evolved and become less expensive, the demand for better, faster, greater capacity and more reliable systems poses significant challenges for facility managers, not only for current needs but also for the future. Organizations have become increasingly dependent on the data that passes through the cabling infrastructure and the loss of even a portion of the system can lead to significant business losses. Supporting these systems must be approached with the same level of consideration and planning as that devoted to other critical facility systems (Piper, 1999a). It is important that facilities have a well-designed power and communication infrastructure, with high capacity and quality.

The level of technology in the building infrastructure is determined by the market, the business and the type of building tenant. User needs range from simple phone and data connections to complex systems incorporating video and other functions. A facility manager must choose a cabling infrastructure design that provides building occupants with voice, data and video communications capabilities that will meet or exceed their business requirements. The design should be universal in scope and flexible enough to support a multi-vendor, multi-technology environment. Before a particular cabling solution can be selected for use in an integrated system, the facility manager should carry out a needs assessment of the organization.

Key criteria in the design life cycle of cabling infrastructure are:

1. *Current business needs*. These range from a phone and a current-technology local area network (LAN) connection in a general office to several strands of fibre optic cable to support high bandwidth applications, along with a coaxial cable to support broadband video distribution and several cables to support multiple PCs in a financial institution.
2. *Occupancy term*. The facility manager must estimate how long the organization is likely to remain in its current building. What is the length of the lease or rental agreement? When there is a turnover in users of the space, it is generally accompanied by a renovation of the space – an ideal time to install new system cables and components.
3. *Future business needs and adoption rates of new technology*. How long are the current users expected to make use of the system? Determine the level of data traffic in existing systems, and how that will be affected by expansion of operations or adoption of new technologies. (Piper, 1999a)

The two basic characteristics of a well-designed and effective power and communication infrastructure are flexibility and reliability. The facility manager must accept the fact that systems change and they must compensate for it by designing flexibility into the system configuration. Bear in mind that while communication electronics and cables may become out of date within five and seven years respectively, cabling pathways can last the entire building life if well designed (Qayoumi, 2000). The higher the rate of changes, moves and additions to the system, the greater the justification for designing in flexibility in order to minimize maintenance and operating costs over the life of the system.

Systems that offer flexibility will have a higher first cost than those that simply meet current needs, but they will often cost less in the long run. New users can be added, existing users can be relocated and new functions can be incorporated in flexible systems without having to make major changes to those systems. In most cases, changes can be made quickly using simple tools and standard connectors with minimal downtime, labour charges and disruption to other tenants. As a rule of thumb, every extra dollar spent in developing the support infrastructure beyond the minimum requirements will save $5 over the life of the system (Piper, 2000).

Building owners and facility managers that have a well-planned and documented distribution and cable management system will find that the cost-effectiveness of the extra steps taken to ensure flexibility will be greatly increased if there are records showing what is installed and where it is located (Kozlowski, 1995; Piper,

2000). Upgrades and changes will become easier and less time consuming as a result.

In the attempt to keep pace with rapid developments in communication technology, the first level of technology that must be addressed is the physical infrastructure of the building or office. To ensure the flexibility of the system design, central facilities must be carefully located, taking into consideration issues such as security, access and protection from water damage. Vertical cores must be designed to allow access at all levels in the facility and must be sized to allow future expansion. Infrastructures require several vertical risers with short horizontal runs that minimize individual cable runs and are suited to zone distribution of power and communication. While this is easy to accommodate in a new construction, in older buildings alternative approaches are to convert smoke towers to risers or give up rentable space. A more common practice is to use existing power and communication closets and pathways to provide the necessary upgrades. Although this can limit flexibility, this approach is permissible as advances in communication technology have decreased the space required for current cables. Another approach is to thread wires and cables through furniture systems or wall panels (Donovan-Wright, 2001). Power, on the other hand, generally requires more space for an upgrade. One should not rely on the availability of more energy-efficient equipment to decrease power needs at the workstation, as these do little more than help slow the increase of power requirements due to increases in office densities and occupants using more equipment.

Managers need a basic understanding of cabling technology; for instance, most systems will only need two types – copper wire and fibre optic cable, and electrical cables are commonly run in conduits or cable trays unless fire-rated. Fibre optic cable is usually recommended to be used between buildings and between telecommunication equipment rooms, while individual desktops are adequately connected to equipment rooms by less expensive copper wire (Piper, 2000). Wires are twisted in pairs or shielded (for example, coaxial cables) to minimize electrical interference. As coaxial cables also have a robust insulating sheath, they are often used for applications that require longer distances and greater bandwidths such as video and Internet (Qayoumi, 2000).

The current or future need for the transmission of voice, data or real time video, large file transfers and high-speed access to the Internet is best met with high-speed copper networks. Installation of multiple outlets provides flexibility to move or rotate staff and common equipment anywhere. The backbone cabling in an office that connects tenants to outside services must be sized to accommodate current and future needs and also address the issues of security and route diversity to avoid failures of the network. The networking hardware must be designed and built to handle the communications needs of the office with a reasonable amount of growth to address the increasing need for speed over the network (Peters et al., 1999).

In place of traditional multiple infrastructure installations, each requiring support to operate, a single, integrated cabling infrastructure can be used to support the communication requirements of all systems, including voice, data, video, life safety and building environmental control. An integrated system is cost-effective in the long run as it is easier and less expensive to relocate and add new users to the system, it can be designed to allow the introduction of new functions and technology without replacement, it offers economies of scale in both mid-size and large facilities, and is easier to maintain and manage. A third type of wireless transmission system is evolving and will play a significant role in the near future (Piper, 1999a).

Besides flexibility, facility managers must guarantee a reliable source of power and communication capacity to safeguard functionality and remain competitive. A sound infrastructure is integral to reliable LAN and voice communication, avoiding downtime, protecting against surges, noise, harmonics and other power variations, and providing additional capacity to building occupants when required. It should be accounted for as part of an organization's strategic planning.

To cater for these requirements there are several approaches. Depending on the purity of the power feeding this equipment, it may be necessary to install surge protectors, power conditioners or uninterruptible power supply (UPS) systems. The greater the level of interference on the system, the higher the level of power conditioning that will be required – a decision that must be made early in the project's design phase (Piper, 1999a). It is recommended to oversize all the power, communication and heating, ventilation and air conditioning (HVAC) systems with either a supplementary redundant power and communication system with back-up power to the building's base systems or a dedicated stand-alone system. This would require reconfiguring the system so there are dedicated lines and risers. Another course is to use two power feeds and a UPS system with an emergency generator and battery back-up. In case of loss of power, the batteries discharge to provide a buffer period in which to make the transition to the generator, which provides the emergency power. In a rotary UPS system, a flywheel carries the load for a short while to reduce the stress on (and therefore extend the life of) the batteries or obviate the need for them at all in momentary outages (Kitco, 2001).

Building owners and facility managers must take control of their building's voice and data infrastructure and improve the system's reliability, the installation process, repair and maintenance operations, and the ability to plan and budget upgrades. Carrying out an inventory of the voice and data wiring system will establish what is available in the risers to meet current and future needs. It will uncover any inefficiencies in the system such as abandoned or unidentified cables, if data transmission speeds are adequate, or if extra telephones lines exist. Documentation of the voice and data network information can be either in a simple spreadsheet or using specific database programs that enable telecommunication system managers to document the components, network and cabling of any size system. The documentation of a building's telecommunication and data infrastructure will help facility managers improve their ability to plan and budget for capital improvements, market their property, and provide reliable voice and data service for tenants or resident employees (Piepmeier and Costanzo, 1999).

To prepare for an infrastructure upgrade it is best to work collaboratively with related groups, and create space-planning teams who are responsible for the overall co-ordination of services. It is also advisable to stay prepared for possible future upgrades by watching industry trends closely, as upgrading a power and communication infrastructure is a far more difficult task than, for instance, buying new computers.

7.3 Infrastructure management

Infrastructure management is a term that describes the growing technological interconnectedness within organizations. There are trade-offs in just how much centralization should take place regarding hardware, software, communications, Internet, data, benchmarks, business processes, and so forth. Organizations have ascertained that investment in infrastructure management yields business value and better understanding and control of assets (Teicholz, 1999). From the facility manager's perspective, technology is constantly changing, and no one really knows where technology will lead. What is clear is that facility managers must understand the implications of infrastructure management for the organization and anticipate how the traditional functions of space and asset management will be affected by this centralizing force.

Teicholz (1999) states that it is hardly a surprise that organizations are spending more and more on their information technology infrastructure. Not so long ago, this infrastructure encompassed only the hardware and software associated with mainframe computers and the networks connecting terminals to mainframes. In the 1980s, personal computers entered the equation. With the 1990s, information technology (IT) infrastructure has expanded to include everything from laptops, palmtops, intranets, extranets, collaborative websites, mobile phones, pagers and personal digital assistants. Given the nature of progress, it is anachronistic to think facility managers will continue to control their IT infrastructure independently of other business units.

The implications for facility management are profound. It means that facility management data must be linked to real estate, human resources, IT, and all aspects of finance and procurement. It means that facility management technology systems must function as a part of an integrated whole, and that the goals of this technology must be in synch with those of the organization.

7.4 Internet-based technologies

The popularity and rapid expansion of electronic mail (e-mail) use in the workplace is a result of various reasons, including lower cost of delivery compared to traditional methods, the almost immediate delivery to the intended recipient, and the conveyance of an editable document that can be updated and returned electronically by the recipient (Peters et al., 1999). The timely delivery of information aids decision-making, allowing for greater efficiency and, as a result, more competitive business practices. E-mail has become one of the basic requirements for participation in today's electronic workplace and every corporation should seriously consider Internet access and a website, and use e-mail to provide customer service if they don't already do so.

The issue of network security is an important one as telecommuters and other staff with remote access to the corporate network need a safe and secure way to transfer and access information. This can be achieved by a virtual private network (VPN) which is a secure, encrypted link over the Internet allowing subscribers access to their corporate network's applications and files. Subscribers can dial a local phone number to connect to the Internet, then access the information on the assigned web page. While the web page may be open to the public, a system of keys and passwords held by corporate subscribers can be used to limit the degree of access to the various content areas.

A VPN enables organizations to securely extend corporate networks to remote offices, mobile workers and major business partners. It also incorporates various levels of encryption, security and permission to distinguish the casual visitor from an employee or subscriber, and to secure transfer of data/information. Dial-up modem access has allowed a mobile workforce to access corporate information from client sites, and as this technology matures, high-speed Internet services coupled with VPN and firewall technologies will make the home office much more effective (Peters et al., 1999).

The Internet, being the most efficient way to communicate and share data with staff and contractors, has also changed the way facilities are managed. Facilities data, project information, work orders, and any other computer-fed information in the company is collected from the appropriate source (including sources outside the company via an extranet). The documents are stored in a server which can be owned by the company or managed by a third party, and then delivered to any authorized entity through a secure intranet or extranet. The Internet (and corporate intranets) can be applied to facility management with off-the-shelf packages that allow anyone with appropriate access to use their browser (i.e., Netscape, Microsoft Internet Explorer) to view data developed by others. Unless there is a need to modify the data, there is no need for anyone to have the software except for the data generator.

Internet technology has an impact on three major areas of facility management:

1. *Internal communication within the facilities department and the company.* Access to facilities or non-facilities management data (telecommunications, human resources, etc.) can be provided through a secure corporate intranet. The facility manager's task is to blend the technology with existing computer-assisted/integrated facility management (CAFM/CIFM) systems.
2. *Project management.* The interactive exchange of project management information can be carried out by users (including those outside the company) tracking a project through the corporate intranet regardless of complexity and the number and types of people managing it. All data can be accessed by those with appropriate authorization.
3. *Client-consultant relationships.* Rather than clients managing their automated systems in-house (after initial implementation by a CAFM/CIFM vendor), Internet technology allows most of the day-to-day management to be dealt with by the consulting firm, freeing the facility manager to concentrate on other issues (Kimmel, 1997).

The benefits of Internet technology include effortless communication and instant access to information, instant awareness of a project's relationship to a proposed facility action, and the ability to show corporate management the outcome of various facility scenarios. Basically the facility manager generates a document through a CAFM/CIFM software package and saves it in a format, such as HTML, so that others on the Internet or corporate intranet can view or print it regardless of whether or not they have any experience with the CAFM/CIFM software. Colleagues who aren't using a CAD package can view and print drawings in either a web browser or as an active, embedded object in standard Windows software programs (for example, spreadsheets, slideshow presentations). Responses or requests can be e-mailed back to the facility manager.

In the area of automation providers, two approaches exist. One approach is where CAFM/CIFM vendors develop products that work over corporate intranets, and the other is a process-based approach where each company starts with a server (usually run by the consulting firm, not the facility manager) and sets up a secure, customized area for each client on the company's own server. The consulting firm makes everything work from a technical perspective, leaving the data entry to the client. These Internet-based systems can take advantage of existing infrastructures and databases already used by the organization. Data maintenance can be performed using standard off-the-shelf CAFM/CIFM systems as an outsource agent, importing other associated client data, adding value with new features unavailable in current CAFM/CIFM systems, and then feeding that information back to the users. An advantage of the Web and a workflow-based approach is that it builds data as a by-product of doing the job (Kimmel, 1998).

The major facility management disciplines are affected by web-based technologies in various ways:

1. *Strategic facilities management*. Facility managers are able to run multiple scenarios that can be instantly accessed by corporate management.
2. *Space management, space planning and furniture management*. Drawings can be viewed with zooming and red-lining functions by anyone; and users can have access to contractors through corporate extranets without need for CAD software or training for those with view-only or red-lining needs. Management and staff can have access to space management, furniture, fixture and equipment data and furniture moves can be requested and planned.
3. *Real property management*. Users can share and access property and lease data.
4. *Maintenance and work management*. Users can process requests such as moves and room temperature changes from clients, and staff have instant access to equipment histories.
5. *Project management*. Users can track requests and project status so they can be aware of projects that may be affected by proposed work.
6. *Other applications*. Facility management procedure manuals and reports can be kept current and distributed without data re-entry. (Kimmel, 1997)

Costs for Internet-based CAFM/CIFM approaches could be 20 to 30% higher than with traditional implementation, however they provide tremendous paybacks – for example, processes that could normally have over fifty steps can be reduced to a four-step data entry process with no re-entry. In addition, a corporate intranet creates a better support system for managing facilities more effectively and strategically with resultant productivity gains. Facility managers can increase their efficiency by managing a construction project or renovation project over the Internet. On a single website, architects, general contractors and subcontractors can review construction documents, notify each other of change orders and update a project's design. In the design stages of a project, product specifications and drawings can be reviewed online. Facility managers can also review properties for sale, view transaction documents and even buy a building (Anonymous, 1999).

The benefits of such technology include:

1. The time saved by management and staff not having to request information from others.
2. Maximum utilization and minimal time wasted by each resource, i.e., facility managers will focus more on facility management, while information systems staff focus on systems maintenance.
3. Reduction of training time to learn new facility management software.
4. Minimal need for specialized expertise to maintain the system if implemented through a consulting firm.

References and bibliography

Alexander, K. (1996). *Facilities Management: Theory and Practice*. E. & F.N. Spon.

Anonymous (1999). 'NetWatch Internet, Web Shaping Nature of Facilities Industry' . *Building Operating Management*, October.
<http://www.facilitiesnet.com/fn/NS/NS3b9jn.html>

Archibus (1999). 'ARCHIBUS/FM, The #1 CAFM Solution WorldWide'. Boston.
<http://www.archibus.com>

Atkin, B. and Brooks, A. (2000). *Total Facilities Management*. Blackwell Science.

Barrett, P. (1995). *Facilities Management: Towards Best Practice*. Blackwell Science.

Cotts, D. G. (1999). *The Facility Management Handbook* (2nd edition). AMACOM.

Damiani, A. S. (1998). *Moving up the Organization in Facilities Management: Proven Strategies to Increase Productivity in your Workforce*. Scitech Publishing.

Donovan-Wright, M. A. (2001). 'Plugged into Change: In high-tech, high-churn offices, flexibility is the hallmark of an effective wiring/cabling infrastructure'. *Building Operating Management*, April.
<http://www.facilitiesnet.com/fn/NS/NS3b1da.html>

Finneran, M. (1995). 'The Devil in the Details'. *Business Communications Review*, 25(7), July, p. 56.

Guzzo, R. A. and Dickson, M. W. (1996). 'Teams in Organizations: Recent Research on Performance and Effectiveness'. *Annual Review of Psychology*, 47, pp. 307-338.

Hamer, J. M. (1988). *Facility Management Systems*. Van Nostrand Reinhold.

Hesselbein, F., Goldsmith, M. and Beckhard, R. (1997). *The Organization of the Future*. Jossey-Bass.

Kimmel, P. S. (1997). 'Internet-based Packages Debut at A/E/C Systems.' *Facilities Design and Management* 16(8), August, pp. 28-29.

Kimmel, P. S. (1998). 'Internet in the House'. *Facilities Design and Management*, 17(9), September, pp. 60-61.

Kitco, J. (2001). 'High Reliability Power'. *Building Operating Management*, March.
<http://www.facilitiesnet.com/fn/NS/NS3b1cf.html>

Kozlowski, D. (1995). 'Meeting Power and Communication Needs: Flexibility and reliability are hallmarks of a well-designed infrastructure'. *FacilitiesNet online*.
<http://www.facilitiesnet.com/fn/NS/NS3bk5c.html>

McGregor, W. and Then, D. (1999). *Facilities Management and the Business of Space*. Arnold.

Moos, R. H. (1996). 'Understanding Environments: The key to improving social processes and program outcomes'. *American Journal of Community Psychology*, 24, pp. 193-201.

Muller, N. J. (1996). *Network Planning, Procurement and Management*. McGraw-Hill.

Pacanowsky, M. (1995). 'Team Tools for Wicked Problems'. *Organizational Dynamics*, 23, pp. 36-51.

Park, A. (1994). *Facilities Management: An Explanation*. Macmillan.

Peters, K., Romani, P. and George, S. (1999). 'Technology in the Workplace: Changing work processes underlie changing space utilization'. *Office Life, Canada*, September.
<http://www.facilitiesnet.com/fn/NS/NS3o9id.html>

Piepmeier, L. and Costanzo, C. S. (1999). 'Wires, Wires Everywhere: If your cabling system is giving you headaches, it may be time for a close look at where all those wires are going'. *Building Operating Management*, October.
<http://www.facilitiesnet.com/fn/NS/NS3b9jj.html>

Piper, J. (1999a). 'Costs, Benefits and the Bottom Line: A flexible design is the key to minimizing life-cycle costs of the cabling infrastructure'. *Building Operating Management*, March.
<http://www.facilitiesnet.com/fn/NS/NS3b9cc.html>

Piper, J. (1999b). 'Beyond Building Automation: Communication between facility and corporate information systems can improve efficiency'. *Building Operating Management*, August.
<http://www.facilitiesnet.com/fn/NS/NS3b9hh.html>

Piper, J. (2000). 'Facilities' High-Wire Act: Achieving a delicate balance between flexibility and cost-effectiveness in cabling and wiring systems'. *Maintenance Solutions*, June.
<http://www.facilitiesnet.com/fn/NS/NS3m0fc.html>

Price, S. (1997). 'Facilities Planning: A perspective for the Information Age'. *IIE Solutions*, August, pp. 20-22.

Qayoumi, M. A. (2000). 'Wired for Performance: Growing demands for power and capacity raise the stakes for wiring and cabling systems'. *Maintenance Solutions*, July.
<http://www.facilitiesnet.com/fn/NS/NS3m0gd.html>

Rondeau, E. P., Brown, R. K. and Lapides, P. D. (1995). *Facility Management*. John Wiley & Sons.

Rosenblatt, B. (1995). *New Changes in the Office Work Environment: Toward Integrating Architecture, OD, and Information Systems Paradigms*. Ablex Publishing Corporation.

Teicholz, E. (1999). 'FM Data Links to RE, HR, IT, and $$$', *Facilities Design and Management*, 18(7), July, pp. 42-44.

Tompkins, J. A. (1996). *Facilities Planning* (2nd edition). John Wiley & Sons.

van Delinder, T. (1997). 'Creating a Progressive Office Environment'. *Facilities Management Journal*, May/June, pp. 18-22.

Building management systems

8.1 Introduction

Over the past twenty-five years, the designers of heating, ventilation and air conditioning (HVAC) systems have gradually shifted toward the use of digital computers, replacing direct manual control and simple analogue feedback loops such as thermostats. Building management system (BMS) installation is essential to efficiently control heating, ventilating, air conditioning, lighting and security systems in large buildings.

Digital control makes possible more flexible, precise and complex control strategies that in turn can provide significant energy savings. For example, most computer-controlled buildings offer automatic temperature setbacks to reduce energy consumption after working hours and on weekends. Many systems also control lighting to save additional energy (Elrod et al., 1993). Occupancy-based control of lights and computer monitoring and control of the air conditioning systems can save a significant amount of money per year. Direct digital controls which tie each unit into a BMS, which is monitored and operated through a personal computer terminal, minimizes the need for people to check and adjust systems. BMS controls need a four- to eight-week commissioning process, depending on system size and complexity, for fine-tuning (Rospond, 1996).

Building management systems can be understood by an examination of building sensors and controls and materials management information systems, and how this affects energy performance and monitoring and intelligent buildings generally.

8.2 Building controls and sensors

A BMS typically consists of a personal computer-based graphic user interface, modular direct-digital control (DDC) panels, DDC-based variable air volume (VAV) box controllers, microprocessor-based 'gateways' to interface or integrate with other systems, and either a copper or a fibre optic communication network. Control of air handling and refrigeration equipment, cooling towers and terminal boxes is carried out through modular DDC control panels that are distributed throughout the building's mechanical spaces. The use of such 'intelligent' field processing units and distributed control techniques removes processing requirements from the central equipment and reduces centralized functions to historical and alarm reporting and status updates (Pearlman and Cana, 1999).

At the nucleus of the DDC system is the programmable microprocessor, which accepts signals from sensors and initiates control actions according to programmed instructions. Since software, not hardware, initiate control sequences and can accept input from a wide range of sensors, DDC systems are more flexible than earlier systems. Modification of a control action requires only changes in software to initiate more complex control strategies. As all of the system's sensors and controllers are linked to a central computer, equipment operation and temperature set points can be remotely monitored, initiated and changed. Mechanics can tap into the system with portable computers or diagnostic equipment to test a particular controller, reset set points and diagnose system problems, often without having to access equipment in hard-to-reach locations (Piper, 1999a). Remote operators can tie into the system by telephone modem and perform any function normally performed by the central operators from outside the facility. System manufacturers also can use the Internet to perform system maintenance without sending out a technician (Piper, 1999a).

A feature available in DDC systems is distributed intelligence architecture, where each control device on the system contains a small microprocessor that can operate in a stand-alone mode should communications with a central computer be disrupted (Piper, 1999a). When operating in stand-alone mode, the device may not be able to perform all of the functions it could as part of the entire system, since it may be cut off from some sensors, but the device will continue to operate under preset conditions.

To make the most of these systems, facility managers need to make sure the system is installed and functioning properly through testing and evaluation. This will ensure fewer start-up problems, better environmental control, lower energy use and fewer long-term maintenance problems. Maintenance technicians should have all of the information and training they need, including the use of computerized maintenance and diagnostic tools to set up and maintain the system. Maintenance and engineering personnel should also have copies of key documentation; operations and maintenance manuals, control drawings and mechanical as-built drawings. Computer hardware and software documentation will be necessary to trace any control problems in DDC systems (Piper, 1997). Routine and preventative maintenance must be performed on control systems if they are to operate at their full potential. All analogue input and output devices should be tested and recalibrated to correct for drift in sensors and actuators, and digital input and output devices should be manually tested to verify proper operation and reporting. Electronic components in DDC systems can be damaged more easily by voltage variations caused by lightning, heavy equipment operation, brownouts and momentary power dropouts than their pneumatic counterparts (Piper, 1999a). They also can be damaged by heat and moisture – two conditions commonly found in mechanical spaces with sensitive components.

Until recently, HVAC control systems and other proprietary systems of fire safety, building security, telecommunications and maintenance management required their own sets of sensors, microprocessors and data transmission systems, functioned independently and could not share information (Piper, 1998c). Since the introduction of standard protocols, systems can integrate a number of independent automation functions into one comprehensive interoperable system that can share data across functions from any point in the system. Interoperability has made the systems vendor-independent, increasing their capabilities while reducing their cost. The systems offer flexibility and are easier to learn and operate. All data is transmittable over one pathway so PCs connected to the system are able to run preventive, predictive and diagnostic maintenance programs, automatically detecting faults and performance deterioration (Piper, 1997; Piper, 1998c). DDC systems are economical for any size facility due to hardware and software standardization. Upgrading a facility's temperature control system to DDC is a cost-effective means of helping to maintain competitive advantage by offering tenants and occupants improved comfort while minimizing energy costs.

8.2.1 Building temperature control systems

DDC systems are well suited for building HVAC controls applications because they can accept data from a range of sensors and use it to drive both simple and complex control strategies, including those that are impossible to implement with pneumatic systems (Piper, 1997). Managers are now replacing pneumatic relays and controllers in temperature control systems in many existing buildings. DDC systems are more accurate, eliminating such errors as offset (when temperature or air flow rate goes above or below the intended setting), set point overshoot (the slow reaction to a changing temperature or air flow rate, allowing the system to go past the desired set point) and set point drift (the variation of temperature or air flow over a range rather than maintaining a setting). This leads directly to energy conservation, improved indoor air quality and greater customer satisfaction (Piper, 1998a; Piper, 1999b).

8.2.2 Boiler technologies

The most important change in boiler technology is the development of digital controls to monitor and control equipment performance that result in safer operation, enhanced energy efficiency, lower maintenance costs and greatly improved diagnostic capabilities (Piper, 1998a). The systems also allow operators to monitor and control multiple boilers at a remote central location. Digital controls are able to manage the operation of a boiler 10 to 100 times more accurately than pneumatic systems as manual control systems tend to lag actual operating conditions, resulting in errors that increase energy use. Conventional pneumatic control systems typically operate with a minimum of 10% excess air to prevent boiler sooting during sudden load changes, while digital control systems can operate with excess air rates as low as 5%. Each percentage point decrease results in a 0.25% increase in boiler efficiency. Digital control systems for boilers also monitor between 10 and 20 times more operating parameters than conventional pneumatic systems. Tracking the data over time helps establish baseline performance characteristics for the boiler, which allows early detection of slow developing problems. Boiler controls can be integrated with the operation of building management systems, allowing even more efficient operation.

8.2.3 Lighting controls

The benefits of lighting control include energy savings, demand reduction, increased productivity and pollution prevention (Urdinola, 1996). Studies have shown that implementation of good lighting control practices generally results in a decrease in energy use of 25 to 35% or more (Piper, 1999b). Lighting controls provide several functions, including occupancy recognition, scheduling, tuning, daylight harvesting and demand control. Some systems perform only one function while others, typically on an automated basis, perform more than one (Urdinola, 1996). When designing the controls for lighting fixtures in any type of area, strategies must be well thought out to achieve proper coverage, flexibility, comfort and energy savings. It often takes an experienced lighting specialist to design energy-efficient offices that suit employees' needs.

Occupancy sensing devices are automatic switches that control lighting based on recognizing the presence or absence of people. Their primary function is to switch lights off in unoccupied space. The predominant types of technology are passive infrared sensors which detect heat emitted from the human body; ultrasonic sensors that 'see around corners' by transmitting inaudible sound; and hybrid sensors that utilize both technologies to ensure proper device operation (Urdinola, 1996). All sensors have adjustable sensitivity settings and each has the flexibility to be used in a variety of applications. The sensitivity and time delay should be fine-tuned after occupants have moved in (Miller, 1999). If well designed, occupants shouldn't notice that the lights are controlled. One of the major objections to occupancy sensors is that they reduce lamp life (Piper, 1999b). The use of occupancy sensors substantially increases the number of starts per day. Each time the lamp is started, there is a slight deterioration that shortens its life. Studies show that single-person offices are occupied only five to six hours a day, while the lights are left on for as many as fourteen hours (Urdinola, 1996). In such cases, even if lamp life is reduced by 50%, occupancy sensors can easily achieve energy savings of 30 to 50% – more than enough to make up the difference for additional lamps and maintenance costs (Piper, 1999b).

Scheduling allows lights to be activated, extinguished and adjusted according to a predetermined plan. Override switches enable people to work after hours. More advanced microprocessor-based technology has made it possible to design timers to control lights at different times each day of the week, and even take into account the seasons of the year (Urdinola, 1996).

Tuning or dimming adjusts light levels for a specific task or purpose. An employee is able to adjust the light level to their specific needs, increasing productivity, and when tasks change, instead of having to change or move luminaires, dimmer switches meet new needs by adjusting the light output. Controlling the illumination level can lead to significant energy use reduction.

Daylight harvesting involves strategically locating photocells to determine the ambient light level so that free daylight can be used to supplement artificial illumination. A control device adjusts the luminaire output to keep the desired illumination level constant. The amount of energy savings that can be realized through daylighting will be affected by factors such as building shape, orientation and materials, the orientation of work areas in relation to windows, the use of window coverings, and possible increases in the cooling load during summer or in warm climates.

Demand control is a strategy to keep utility demand charges to a minimum by shedding secondary loads during peak periods. Possible sources are lobby lighting systems, perimeter office lighting, and other systems that can be dimmed for short periods without negatively affecting safety or security. An energy management system is one way to control demand and to implement other lighting control strategies.

A lot is at stake in the selection of lighting controls, such as energy savings, worker productivity, error reduction, quality control, employee morale, environmental enhancement, security improvement and the like, so it is important to consult with an expert to get the maximum return on investment.

8.2.4 Smoke detectors

Smoke detectors respond to different kinds of smoke. For example, ionization-type detectors are most responsive to flaming (less smoky) fire conditions, while photoelectric sensors detect smouldering or visible smoke fires (Schumacher, 1997). Addressable fire alarm systems are able to pinpoint the specific detector from which an alarm originates. This allows a more accurate response to an incident but can increase the degree of complexity in software programming that may not be warranted in all cases. Analogue or intelligent sensors are further able to detect how much smoke or heat there is, in contrast to conventional detectors, which only indicate that there is a fire present. They can also analyse and store the detector's performance and sensitivity data. These devices can actually compensate for dirt or other contaminant build-up over time, and communicate to the system's control panel that maintenance is necessary before false alarms become a problem (Hauf, 1996). Multi-sensor applications process signals from two or more types of detectors to provide quicker responses to different types of smoke or fires than would single technology detection. Some detectors have microprocessors which can differentiate signals from non-combustion particles, such as dust or vapours, thus reducing false alarms (Schumacher, 1997).

To take full advantage of these systems, the appropriate technology must be chosen for the fire detection application. Environmental factors, such as air flow and quality, ambient temperature and humidity, electromagnetic or radio frequency interference, and lighting have to be considered as they influence system choice. Decisions on the appropriate types of detection technology should be made on an area-by-area basis, and should be reviewed when the use of an area changes (Schumacher, 1997).

8.2.5 Demand-controlled ventilation

Demand-controlled ventilation (DCV) monitors indoor air quality, adjusting the quantity of outdoor air to maintain a healthy indoor environment while minimizing energy use through reduced refrigeration/dehumidification needs. Controlling the amount of outside air to be brought into individual rooms or zones at any given time is achieved by a low-cost electronic monitoring device that accurately measures carbon dioxide levels (Maybaum, 1999; Piper, 1999a). The carbon dioxide sensing system can maintain a target per-person ventilation rate as occupancy, density and activity vary throughout the day (Anonymous, 1998). The software controls outdoor air dampers so the system operates optimally only when those spaces are in use. DCV can also be incorporated into systems that have separate ducting for fresh air supply, but as installing an independent fresh air delivery system is fairly costly, the potential benefits of such a system must be carefully considered (Maybaum, 1999).

The sensors use infrared light to monitor carbon dioxide levels and are unaffected by changing humidity, temperature and duct pressures, and only require simple calibration every five years. Sensors that monitor other contaminants, such as volatile organic compounds or carbon monoxide, which is useful in parking ventilation, are also available (Maybaum, 1999). Generally, duct-mounted sensors work best for single zones that have a uniform occupant density and activity, while wall-mounted sensors more closely represent real-time conditions. High efficiency filters are recommended to deal with poor quality outdoor air. Having extra space built into air-handling units is also wise in case additional or higher-efficiency filters need to be added if the outdoor air quality deteriorates or if, for example, the owner wants to attract tenants with very strict air quality requirements. The small additional cost of building additional space may be easily justified if the business environment changes or if it gives an owner an edge in attracting prospective tenants (Maybaum, 1999).

8.2.6 Other control systems

Other sensors can be integrated into a building management system to provide additional automatic functionality, such as closing blinds to compensate for changing natural light levels, lowering light levels while a workstation is in use, and adjusting ventilation depending on whether doors are opened or closed (Elrod et al., 1993). Security systems, including integrated video badging, access control, intercom and closed-circuit television (CCTV) monitoring, can be integrated with traffic, parking, fire alarm and lighting systems, as well as the DDC system to perform various functions. For example, HVAC and lighting levels can be controlled upon authorized access and by CCTV activation. By the time the person reaches their office, the lights are on and the room temperature is approaching the desired level. Card-reader controlled doors can be opened to provide emergency egress, and CCTV can be used to provide a functional alarm assessment (Pearlman and Cana, 1999). Fire, security and environmental systems can be linked to a paging system so that an appropriate manager is notified of a critical event, such as smoke detected at a remote site or of temperature changes in a critical area (Tatum, 1998).

8.2.7 Integrated automation

In advanced building management systems, all major systems are integrated, including HVAC, lighting, fire and security, as well as the communication technologies for voice, data and office automation. A single network featuring a fibre-optic backbone can simultaneously handle voice, data, video, security and building control systems (Tatum, 1998). The benefits of such a system include central monitoring and control of mechanical systems, which enhances comfort, energy efficiency, security and cost-effectiveness. In new construction, integrated systems typically cost 10 to 15% less to install than independent security and fire safety systems, mainly due to the use of a common cabling system. If, however, existing independent systems are being replaced, the cost of replacing them with an integrated system rather than independent systems is significantly higher because of the need to replace the existing cabling. The use of a common cabling system and user interface results in typically 10% lower operating and maintenance costs than for independent systems. The use of a single, standard operating system also reduces personnel training time (Piper, 1999c). Integrated systems allow facility managers to gather accurate energy profiles of the buildings, tenant billing or power quality management. The savings produced by energy management can turn the automation system from a business expense into a business tool (Tatum, 1998).

Using advanced controls, facility managers can tailor solutions to a particular building's needs (Tatum, 1998). For example, light sensors used to calculate the amount of electric light needed to supplement daylight levels can be used to predict heating and cooling routines for different building zones in an integrated system. The most sophisticated integrated HVAC systems include environmentally responsive workstations (ERW), which employ a panel of controls on workers' desks (Anonymous, 1997). Employees in open-office areas can individually control temperature and airflow within their spaces through vents and radiant heaters built into their furniture, and control task lighting and mute ambient noise around their workstations. Integrated HVAC systems also allow monitors to control condensation levels in different areas of a building, which is critical to businesses that have humidity-sensitive equipment (Anonymous, 1997). Fire and life safety notification can be used as a back-up for de-energizing supply and exhaust air systems through air handler control routines.

Linking fire and life safety systems with security management systems allows facility managers to cut operating costs and reduce response times to critical fire and life-safety events. Integrating the security and fire safety systems also eliminates the need to maintain duplicate databases. It allows an accurate logging of all critical events using an online history database so that a facility manager can play detective, for example finding the cause of a suspicious fire or recent false alarm. Building security too is a function made considerably more efficient by integrated monitoring systems. Along with video and audio devices that can monitor events throughout a building, access cards carried by employees can be read by a central computer system, and monitor their movements throughout the day (Anonymous, 1997). A drawback of integrated systems is, in the event of a failure, that all elements of the system can be disabled. In an independent system the impact of a failure is limited to that single system. Improved system checking and diagnostic tools can help to reduce the chance of a system-wide failure, but not completely eliminate it (Piper, 1999c).

8.3 Materials management information systems

A materials management information system (MMIS) collects information on the attributes and condition of components of an infrastructure and enters them into a database. It identifies and prioritizes repairs, schedules preventive maintenance, and captures costs of labour, equipment and materials. MMISs are changing so rapidly that selecting a product is an intricate procedure. The decision process should include the following:

1. Obtain as many impartial, first-rate opinions as possible.
2. Keep up with the market by reading trade publications with information on prices, number of installations, capabilities, etc.
3. Form a meeting with everyone who will have access to the system and look at every system on the market.
4. Narrow the list of vendors. Aim for five or ten of the best, most applicable, systems.
5. Ask vendors to come in and perform demonstrations, or if time permits, to visit installed sites.
6. Once the necessary capabilities of the system have been determined, set up the criteria for the system and ask the vendors many questions.
7. Before purchasing a particular MMIS, look to the future. Include features that in all probability will be useful in the future (Anonymous, 1996).

8.4 Energy performance and management

Most facilities today can reduce energy use by 25–50% by changing how facilities are operated, by making existing energy-using systems operate more efficiently, and/or by making significant investments in technology (Piper, 1998b). Energy management programs that use all three of these methods will achieve the greatest results. Simple energy conservation measures, such as optimization, waste reduction and efficient operation, can be implemented to initiate the energy conservation program at any time with minimal investment (Irrinki, 1997). Energy optimization is more than energy conservation or energy efficiency. Energy conservation simply reduces energy without regard to productivity, while energy efficiency goes a step further by reducing energy while simultaneously maintaining or improving productivity. Energy optimization combines both conservation and efficiency with operational procedures, not only to generate long-term savings, but also to increase productivity and sales (Tatum, 1999a). In order to develop a program to improve energy performance, facility managers should make sure that they have access to information on the facility's energy use, and have the ability to monitor and control that energy use.

8.4.1 Energy auditing

The first step towards controlling a building's energy use is completing a comprehensive energy audit. An audit should include a rigorous analysis of operating and maintenance procedures to identify and quantify how energy is being utilized and distributed. An audit should be completed or updated every three to five years or whenever major changes occur, such as staff resizings, renovations or new ownership (Sieben, 1999). The energy audit should identify savings opportunities that can be readily implemented at little or no cost, and will generate savings over a twelve to eighteen month period to recover the initial cost of the audit. The audit may uncover maintenance procedures or operations that can be simplified or revised, and identify capital improvements, such as lighting and chiller upgrades, that will save the organization money over the long term. If your organization owns or manages multiple facilities, developing an overall portfolio energy review will help to make informed decisions as to which properties offer the greatest energy-saving opportunities. Portfolio reviews provide senior management with a better understanding of expected savings and can lead to better decision-making. Energy management then becomes a strategic initiative within the organization (Sieben, 1999).

To determine a facility's potential energy savings, the existing energy use levels in the facility should be benchmarked against the energy requirements of similar facilities (Piper, 1998b). For sensible comparison between facilities, an energy use index (EUI) needs to be employed. This figure is the total annual energy use of a facility divided by its total occupied or conditioned area. As this figure doesn't account for climate changes or for variations in use, careful selection of similar facilities is required for successful benchmarking. Sources for EUI data are from other facility managers, or from government publications, trade association reports and energy management handbooks that have average values for a range of building types of different age, size, climate, occupancy schedules, construction and energy source.

8.4.2 Energy management programs

Once a facility's energy savings potential has been determined, the next step is to target areas for optimization. To deliver energy efficiency gains, facility managers need a program that identifies opportunities for improving performance through proper maintenance of energy-using systems, application of new technologies to existing systems and replacement of inefficient systems with new, high-efficiency ones. Areas that offer the greatest potential for energy efficiency are the major energy-using systems of HVAC and lights, and should be targeted first (Piper, 1998b). Before considering major replacement or upgrade programs, make certain the existing systems operate efficiently and effectively. Facility managers should examine the maintenance records of the systems, checking for significant maintenance problems. Retrofitting efficiently operating existing systems with direct digital controls, variable-speed drives and automatic combustion controls can provide additional energy savings. If, however, systems are approaching the end of their performance life or if they cause major maintenance problems, it may be more cost-effective to replace them than to overhaul them.

For many facility managers wanting to upgrade ageing equipment and install energy-efficient technologies, finance becomes an issue. A range of financing options for viable energy upgrades exist (Tatum, 1999a). These include straight financing (which means using capital funds), performance contracts from energy service providers and companies (which guarantee savings to help pay for the equipment upgrades), capital leases for energy upgrades and operating leases (with or without performance guarantees), and outsourcing energy services (where the vendor owns, operates and maintains the energy system, and the facility buys heat and light).

Whether using an outside consultant or carrying out energy upgrades in-house, upper management needs to be kept informed with progress reports. Employees and building occupants will also need to be kept informed, explaining the economic reasons and other benefits involved in the energy upgrades. Continual evaluation of implemented measures and reporting of results is required to verify savings are as anticipated and make sure occupants are satisfied. The monitoring of energy use is crucial to make sure energy continues to be used efficiently and to avoid what the experts call 'energy creep' (Tatum, 1999a).

8.4.3 Energy monitoring

One of the most important and time-consuming tasks that must be completed in energy management is gaining an understanding of how, when and where a facility uses electricity (Vergetis-Lundin, 1998). While utility bills provide data on how much electricity was used and at what times, they do not provide the level of detail required to hone the energy performance of a facility. Performance monitoring is essential for the control and evaluation of an energy conservation program (Irrinki, 1997). Effective measuring methods are necessary to control, evaluate and manage efforts to conserve energy. Manual reporting or feedback from the meters of an energy management system can achieve this. The primary electrical meters of an energy management system send data, sometimes in real time, to a central computer where it is processed and presented for evaluation in a number of different formats, including tabulated listings and spreadsheets (Vergetis-Lundin, 1998).

Energy management systems offer the facility manager a very powerful tool that can be used to help oversee both the costs and the operation of a facility. Energy management systems are cost-effective in commercial, institutional and industrial facilities of all sizes, and can reduce energy use by 30% or more with typical paybacks of two to five years (Piper, 1996). Apart from reducing energy use, they are used to monitor and control HVAC systems in remote locations, allowing changes to be made by a central operator, as well as receive early warnings of building system malfunctions, often before occupants are aware there is a problem. Through the use of direct digital control strategies, they provide more flexibility and ten times more accurate space temperature and humidity control than can be achieved with conventional control systems. They also provide control in areas such as security and fire safety (Piper, 1995; Piper, 1996).

The energy management system can help to control the use of electricity by performing three tasks: tracking electrical demand, rescheduling the operation of electrical equipment and curtailing the use of electrical equipment. Data on energy use will allow the facility manager to plot a load profile for the facility on a daily, weekly, seasonal or annual basis. A facility should aim for a load profile that limits peak demand and is as flat as possible as electricity provider rates will be more favourable. With an energy management system that tracks the electrical energy use of the facility, target levels can be established in the system to automatically trigger an alarm when electrical demand reaches a predetermined level. Electrical demand can then be reduced by performing a process of load shifting and/or curtailment. Electrical load shifting by scheduled operation involves operating certain electrical loads, such as thermal storage systems, only during off-peak periods. Load shifting by alternative sources involves operating equipment that uses a fuel other than electricity during hours when electrical rates are high or at peak electrical demand. Load curtailment automatically turns off electrical loads when they are not needed, or temporarily turns off or throttles back electrical equipment, such as building air handlers, chillers, lighting systems and freezers, on a rotating basis to lower the demand for electricity.

Finding the right energy management system best suited for a facility starts with an understanding of the capabilities of the systems and continues through the installation process to system operation (Piper, 1995). It is important to understand that no energy management system can overcome shortcomings due to maintenance neglect or design problems. These must be corrected before installing an energy management system. The energy management and automation needs of a facility must then be determined, and a system that most closely matches those needs should be selected. Regardless of the requirements of a facility, there are several features that should be included in any energy management system installation. Energy management systems should have an expansion capability (at least double in size), guaranteed vendor support (parts availability, software upgrades, emergency service, etc.), a non-proprietary communication system (standardized interfaces allow transfer of data to other applications, such as maintenance programs, databases, word processors and spreadsheets), and the capability of generating customized reports (Piper, 1996).

To get the most value from the investment, time must be devoted to three important tasks: staffing, maintenance and documentation (Piper, 1995). Operating staff must be trained on how to use the system interface. As automated as energy management systems are, adequate number of staff will have to establish and maintain the operating schedules for system-controlled equipment. The facility manager must ensure that the system performs effectively over its life, with upgrades to hardware and software, and addition of new monitoring and control points. The larger the system, the greater the need for accurate documentation. Without it, system maintenance becomes difficult, particularly as changes and upgrades are made to the original system.

8.4.4 Energy information and analysis software

Energy information and analysis software can help monitor and manage consumption and costs by delivering information on system performance, including changes that may result in higher operating costs, discomfort and breakdowns. Systems incorporate comprehensive energy management into distribution monitoring by including information obtained from meters, circuit breakers, protective relays and motor starters, and some have real-time decision-making and accounting features. Some software allows the facility manager to prioritize energy-efficiency improvements. The software can break down how an energy bill is calculated, allowing facility managers to accurately verify utility bills and track specific charges, and evaluate alternative rate structures. Billing components, such as demand and energy consumption values, can be modified to evaluate future energy usage. Users can often simulate 'what if' scenarios regarding the impact of changes in production and scheduling, as well as to determine the true cost or benefit and perform facility-to-facility comparisons for benchmarking (Vergetis-Lundin, 1998; Vergetis-Lundin, 1999).

8.4.5 Operation and maintenance

In many cases, a thorough maintenance program may be a significant factor in optimizing energy performance, enabling the equipment to operate at design efficiency, resulting in significant reductions in energy usage and increasing the life of equipment. An operation and maintenance manual should provide critical information such as equipment specifications, start/stop/shutdown/standby procedures, operating logs, preventive and routine maintenance procedures and performance details. Other improvements in plant efficiency can be achieved through proper sequencing of process operations, rearranging schedules to utilize process equipment for continuous periods of operations in order to avoid numerous short cycles and minimize preheat losses, and scheduling process operations during off-peak periods in order to avoid peak demand charges (Irrinki, 1997).

Lighting systems offer large, potential energy savings through changes in operations and maintenance. Measures include installing automatic lighting controls in frequently unoccupied spaces, fluorescent lighting system maintenance involving group relamping when lamps reach 70–80% of rated life to reduce labour costs by 90% and improve lighting levels, and replacing existing light sources with more efficient ones (Piper, 1998b). Up to 10% of lighting costs may be saved by simply cleaning fixtures according to the International Association of Lighting Management Companies (Tatum, 1999a). Fixture light output may be improved from 10% in enclosed fixtures in clean environments to more than 60% in open fixtures located in dirty areas. As a result, fewer lighting fixtures can be installed than before and productivity in the workplace may be improved. The impact of lighting has an effect not only on the energy charge, but also on air-conditioning systems. Lighting levels should be tailored to the individual requirements of the task to be performed.

HVAC systems installed to provide comfort cooling or heating are the ideal candidates for energy savings. The maintenance procedures of primary energy-using systems should be checked, for example if boilers, cooling towers and chillers have an effective water treatment program, and if chillers, cooling towers and boilers are inspected and cleaned regularly. Performing required maintenance increases the operating efficiency of these systems, resulting in large potential savings. Distribution losses can be minimized with adequate insulation and short pipe runs of optimum size. Boiler controls should be updated, as automation can increase the efficiency of the system (Irrinki, 1997). Replacement or retrofitting of existing equipment or making changes to the process will also result in energy savings. Modifications can be made as a result of equipment life, equipment reliability, building load changes, building expansion, building occupancy changes, new quality control measures, innovative design concepts, availability of energy-efficient components or changes in the building codes and standards (Irrinki, 1997).

8.4.6 New buildings

Studies show that a building can be designed to use 50 to 70% less energy than a typical building, but this has more to do with how building systems go together than what specific new technologies and energy-efficient equipment are used (Sullivan, 1998a). Dramatic energy gains are possible across a wide range of buildings with good designs (including careful siting, orientation and shape of the building, and window size and orientation) and standard energy-efficient technologies and products (including materials, lamps, ballasts, chillers, motors, variable frequency drives and glazing) at zero or very low additional cost. Energy performance is ultimately determined by the way the buildings function as a whole. Designing individual pieces of the building with all of the other parts in mind is integral to any attempt to produce an energy-efficient building. Therefore architects, engineers, lighting designers, space planners, etc., must co-operate as a team to define needs, identify problems and opportunities and discuss innovative approaches. Ideally, the facility manager should act as the co-ordinator who chooses the team members and sets the team's agenda. The intensive communication required to make the team-design effort succeed generally produces a building that works better, saves money and is more desirable to occupy.

8.5 Intelligent buildings

The advent of microprocessors in building HVAC, security, lighting, power, fire safety and access systems has significantly increased building owners' and managers' control over the operation of their facilities since the days of electromechanical devices, such as the simple thermostat. Configurations of computerized systems have evolved from performing specific, isolated tasks to being able to interface with other systems in the building. They share information and initiate specific control actions based on input from multiple systems (Piper, 1998b). Computer-based control building automation systems predominate in most commercial and industrial buildings, reducing energy costs while improving system performance, operability and reliability.

The once incompatible systems are being integrated through the use of a common communication backbone and control console (Pearlman and Cana, 1999). The most advanced integrated systems link nearly all building functions, including HVAC, security, lighting, audio/visual, fire alarm, lifts and power monitoring through the use of standardized communications protocols and high-speed gateways. It is also possible to integrate select building management systems with telecommunications and office automation systems, so that, for example, air conditioning terminal boxes can be opened through a telephone, or overtime operation can be requested by scheduling an after-hours meeting on a personal calendar. Intelligent building management systems are capable of meeting individual needs, such as human comfort or catering for flexible work hours and patterns, while controlling energy use.

8.5.1 Advantages

When a project is designed and implemented using an intelligent building approach, the whole is greater than the sum of its parts, resulting in a building that provides an organization with the flexibility and adaptability that actually facilitates change. This is particularly crucial for businesses that experience high churn rates (Caloz, 1999). An intelligent building structure will provide the flexibility to create office space and storage, reconfigure computer connections and adjust other support systems to meet limited time and space needs. Intelligent building design strategies allow the development of a support infrastructure to accommodate future technologies. Workplace flexibility can make the difference between occupied and vacant office space (Loerch, 1995).

Today's infrastructure technology and systems provide an effective way to support an organization's business objectives. An integrated intelligent building infrastructure maximizes an investment in technology, especially in information systems, computerized databases and research applications. Sometimes referred to as 'information age buildings', intelligent buildings provide optimal environments for information workers (Caloz, 1999). Tenants should be provided with the technical support necessary to take advantage of an intelligent facility, such as an in-house technical support firm that provides round-the-clock technical service (Loerch, 1995).

With 50% of a building's lifetime costs typically going to operations, however, the most attractive benefit of integrated buildings may be its operating costs (Pearlman and Cana, 1999). Energy consumption is significantly reduced when the operation of lighting and HVAC systems is limited to occupied areas, and out-of-hours use is monitored. Further, the control of multiple systems from a single console requires fewer operators which significantly reduces staff costs. A network of sensors monitoring building and equipment functions will also reduce the need for on-site maintenance staff. If a malfunction is detected, sensors automatically dial out to contractors, the owner or an off-site maintenance manager, and the problem may even be corrected over the phone (Loerch, 1995).

The intelligent building approach avoids the inevitable mess that results in a wiring closet when many layers of wires are pulled, connected, disconnected and reconnected as technology changes and people move. If one contractor installs the wire for all systems on a common infrastructure, less cable is required. The architectural infrastructure is optimized and labour is used efficiently (Caloz, 1999). The core of an intelligent building is typically the HVAC-centred building management system, since it is distributed throughout the building and can be easily linked to other systems (Pearlman and Cana, 1999). Linkage is made easier through the use of variable frequency drive motors and other controllable components that link directly into the building management system. While electronic components (such as telecommunications, monitoring and wiring) are more commonly associated with smart buildings, smarter HVAC monitoring can also be a part of the overall intelligent building infrastructure to make a facility operate more efficiently (Loerch, 1995).

8.5.2 Communication pathways

Caloz (1999) suggests that intelligent building systems are generally grouped into two distinct categories: building operations (environmental controls, security, lighting control, fire/life safety, power monitoring and transport) and user operations (data, voice and video). Integration can range from merging all the systems (using the same command protocol, wiring infrastructure, sensors, etc.) to systems interoperability (where systems share data input from sensors but operate independently over unique wiring backbones, using proprietary protocols and relying on different command hierarchies). Intelligent building systems can exchange information over a dedicated backbone, such as a copper or fibre optic cable, or can utilize a building's existing voice and data cable infrastructure. By providing a dedicated cable, communications problems such as collisions, interference and heavy traffic periods are minimized, however this approach does increase the installation costs and the need for in-house personnel to maintain the communications loop. Using a common communications backbone to support a single integrated system results in a more flexible network that can be easily modified to support changes in building use. Part of the building network can be used for communications by adding standardized communication cards, thereby enabling the systems to reside on the Internet. This eliminates the cost of installing a new communications riser, and existing communications closets can be used. Communications problems and network administration can be handled by in-house professionals.

8.5.3 Interoperability

Interoperability is the connectedness of systems from different manufacturers and of different types to share information with each other without losing any of their independent function capabilities (Tatum, 1999b). Interoperability provides benefits to building owners and managers, by reducing installation costs (due to the use of shared wiring), operator training requirements and operating costs (Piper, 1998c). System operation within the facility is better co-ordinated, and with all of the information generated by the building's systems available at one source, information is more readily available to make operating decisions and initiate control actions anywhere in the system. Using a common language is more practical than using microprocessor-based gateways as it allows communication between products and systems from different suppliers on the same communication media without having to use many different and expensive translators to cope with the multitude of languages in a building management system (Caloz, 1999).

8.5.4 The intelligent building approach

Traditionally, commercial and institutional buildings were built to provide basic services that met generalized, minimal codes and standards. In contrast, the intelligent building is a flexible and adaptable building that helps users meet their goals and supports occupant functions. A holistic approach is required in its design, involving the technology designers (information systems, security and building automation) from its inception (Caloz, 1999). An intelligent building is created through the coupling of technology and innovative design strategies. Programming the intelligent building infrastructure requires identifying the goals of the owner and occupiers of the building, their needs with respect to intelligent building and user operations, the organization's threats and vulnerabilities and its assets and how they are protected. The engineer will need to look at the organization's future requirements for voice and data on the desktop and what kind of backbone and capacity support will be needed to do that. Realizing an intelligent building concept requires someone in the organization to uphold it at every stage. When the project goes out to tender, the system should be packaged so that all the specialities (voice, data, fire/life safety, etc.) are put under one prime intelligent building systems contractor. This assures contractor co-ordination.

8.5.5 Retrofitting

Intelligent building principles are not restricted to new construction projects. Although building intelligence into a facility from the ground up is the preferred method, intelligent system retrofits can be a highly viable option as most facilities have enough existing wiring to implement most new technologies (Loerch, 1995). If a building has existing building operating equipment and systems, new monitoring and control technology can be integrated on existing telephone lines. Integrated building systems can also be added during any major voice-data infrastructure upgrade. Integration should provide a foundation for future growth and implementation of new technology opportunities, while accommodating present-day requirements. A retrofit requires planners to accommodate the existing infrastructure and to pre-planning the network layout (Loerch, 1995). Measures should be taken to avoid taxing the wiring or weakening the line signal.

8.5.6 Implementation

It is widely acknowledged that integrating automated building systems at the software level creates gains in a wide range of areas, including greater energy savings, added flexibility, better control, a more productive facility staff and happier occupants. The decision as to the viability of implementation in a facility should be made with the following considerations (Sullivan, 1998b):

1. *Use*. Determine which capabilities will actually pay off for an organization. For example, does the facility staff have the resources to analyse the large amount of

information generated by an integrated system, or does the organization have an accounting system that takes advantage of the ability to bill different departments their energy use?

2. *Cost*. Integrating system software is cheaper than tying them together with hardware interfaces, and is easier to maintain and change. To take advantage of that first-cost advantage, facility managers should keep control-system integration needs in mind when evaluating tender packages for individual building systems. It might be more cost-effective to select a higher cost bid if an integrated package is available, than linking cheaper systems with hardware later.
3. *Proven reliability*. Facility managers should find out if specific systems have ever been integrated before, and ask for tests to prove that integration will do what it is supposed to. Anything can be made to work in the factory, so specifying demonstration mock-ups and having systems tested independently is advisable.
4. *Acceptance*. Do not overlook the importance of getting the new system accepted by the operations staff by asking for their input – advantages and disadvantages may be identified that may otherwise have been missed.
5. *Ease of use*. One of the selling points of integrated systems is simplicity. Avoid the possibility of unwarranted, added complexity by identifying what facility needs are to be met by integration.
6. *Service*. When a group of building systems is integrated, a malfunction in one subsystem could cause everything to crash. Therefore it is important that facility managers ensure problems will be handled immediately. Assess the availability of local service, and make sure that they understand what the service is and how it will affect the overall system.
7. *Responsibility*. An integrated system can provide a single point of responsibility for system functionality, either in the form of a single-vendor control system (who also typically provides service and upgrades) or a nominated system integrator.
8. *Fire safety*. In an integrated system, fire alarms have to be given top priority. With an integrated system that controls fire safety, HVAC and security, the HVAC controls could keep the entire system, including the fire safety portion, from being operational. Reservations about linking security and fire alarm systems to building management systems because of the fear of security breaches or life safety problems from system malfunctions are diminished with the availability of very reliable software and hardware safeguards that ensure trouble-free linkage (Pearlman and Cana, 1999).

References and bibliography

Alexander, K. (1996). *Facilities Management: Theory and Practice*. E. & F.N. Spon.

Anonymous (1996). 'Buying an MMIS: An intricate procedure'. *Health Management Technology*, 17(12), December, p. 17.

Anonymous (1997). 'The Case for Building Automation: The rise of microprocessors fuels the spread of systems to enhance safety and comfort'. *Maintenance Solutions*. <http://www.facilitiesnet.com/fn/NS/NS3m7kd.html>

Anonymous (1998). 'Ventilate Spaces on Demand using Carbon-dioxide Sensors'. *Buildings*, 92(2), February, p. 24.

Atkin, B. and Brooks, A. (2000). *Total Facilities Management*. Blackwell Science.

Barrett, P. (1995). *Facilities Management: Towards Best Practice*. Blackwell Science.

Caloz, J. W. (1999). 'Real Intelligence: Smart buildings fizzled in the '80s, but 21st century business needs have given the idea new life'. *Building Operating Management*, May. <http://www.facilitiesnet.com/fn/NS/NS3b9ed.html>

Cotts, D. G. (1999). *The Facility Management Handbook* (2nd edition). AMACOM.

Elrod, S., Hall, G., Costanza, R., Dixon, M. and des Rivieres, J. (1993). 'Responsive Office Environments', *Communications of the ACM*, 36(7), July, pp. 84-85.

Hamer, J. M. (1988). *Facility Management Systems*. Van Nostrand Reinhold.

Hauf, J. E. (1996). 'Product Trends: Fire safety systems'. *FacilitiesNet.* <http://www.facilitiesnet.com/fn/NS/NS3b46f.html>

Hesselbein, F., Goldsmith, M. and Beckhard, R. (1997). *The Organization of the Future*. Jossey-Bass.

Irrinki, N. (1997). 'Generate Savings with Energy Conservation Program: Implement an energy conservation program only when the savings can offset the implementation cost over a period of time'. *AFE Facilities Engineering Journal*, September/October. <http://www.facilitiesnet.com/fn/NS/NS3a7ic.html>

Loerch, W. K. (1995). 'Applications'. *Building Operating Management*, August. <http://www.facilitiesnet.com/fn/NS/NS3appli.html>

Maybaum, M. W. (1999). 'A Breath of Fresh Air: The air side of HVAC systems offers overlooked opportunities to reduce costs and improve IAQ'. *Building Operating Management*, January. <http://www.facilitiesnet.com/fn/NS/NS3b9aa.html>

McGregor, W. and Then, D. (1999). *Facilities Management and the Business of Space*. Arnold.

Miller, N. (1999). 'Shedding Light on Productivity: Proper lighting may well reduce eyestrain, improve attitudes and lead to increased productivity'. *Building Operating Management*, August. <http://www.facilitiesnet.com/fn/NS/NS3b9hi.html>

Park, A. (1994). *Facilities Management: An Explanation*. Macmillan.

Pearlman, A. and Cana, O. (1999). 'A Smoother Road: Yesterday's obstacles to integration shouldn't block today's plans'. *Building Operating Management*, August. <http://www.facilitiesnet.com/fn/NS/NS3b9hg.html>

Piper, J. A. (1995). 'Riding Herd on Energy Costs: An EMS can help do the job, but only if it has been selected with the needs of a specific facility in mind'. FacilitiesNet. <http://www.facilitiesnet.com/fn/NS/NS3bj5a.html>

Piper, J. A. (1996). 'Product Trends: Energy management systems'. *FacilitiesNet.* <http://www.facilitiesnet.com/fn/NS/NS3b86i.html>

Piper, J. A. (1997). 'Control Considerations: Getting the most from control systems means understanding their power and flexibility'. *Maintenance Solutions*. <http://www.facilitiesnet.com/fn/NS/NS3m7ic.html>

Piper, J. A. (1998a) Cutting-Edge Performance: Facilities reap benefits from advances in HVAC technology'. *Maintenance Solutions*, April. <http://www.facilitiesnet.com/fn/NS/NS3m8dk.html>

Piper, J. A. (1998b). 'Energy Efficiency: Energy Management Opportunities'. *Maintenance Solutions*, June. <http://www.facilitiesnet.com/fn/NS/NS3m8fk.html>

Piper, J. A. (1998c). 'Interoperability Update: Facilities benefit as systems become less proprietary'. *Maintenance Solutions*, January. <http://www.facilitiesnet.com/fn/NS/NS3m8ak.html>

Piper, J. A. (1999a). 'Controls: Riding the Digital Wave'. *Maintenance Solutions*, September. <http://www.facilitiesnet.com/fn/NS/NS3m9ii.html>

Piper, J. A. (1999b). 'Seeing the Light: Does longer lamp life mean lower costs?'. *Building Operating Management*, February. <http:// www.facilitiesnet.com/fn/NS/NS3b9ba.html>

Piper, J. A. (1999c). 'Time for the Next Step?: Technology makes it possible to integrate security and fire safety systems, but before you make the move, be sure you know what you want to achieve'. *Building Operating Management*, November. <http://www.facilitiesnet.com/fn/NS/NS3b9kf.html>

Price, S. (1997). 'Facilities Planning: A Perspective for the Information Age'. *IIE Solutions*, August, pp. 20-22.

Rondeau, E. P., Brown, R. K. and Lapides, P. D. (1995). *Facility Management*. John Wiley & Sons.

Rosenblatt, B. (1995). *New Changes in the Office Work Environment: Toward Integrating Architecture, OD, and Information Systems Paradigms*. Ablex Publishing Corporation.

Rospond, K. M. (1996). 'Barneys Dresses up Building for Success'. *Consulting-Specifying Engineer*, 19(1), January, pp. 24-28.

Salvendy, G. (1997). *Handbook of Human Factors and Ergonomics* (2nd edition). John Wiley & Sons.

Schulze, P. C. (1999). *Measures of Environmental Performance and Ecosystem Condition*. National Academy Press, Washington.

Schumacher, J. (1997). 'Fight Fire with Technology: System advances give facilities more power to detect fires and save lives'. *Maintenance Solutions*, May. <http://www.facilitiesnet.com/fn/NS/NS2m7ec.html>

Sieben, C. R. (1999). 'The Path Best Travelled: A mountain of engineering reports won't get an energy upgrade approved, but clear communication and presentation to the right people could'. *Energy Decisions*, November. <http://www.facilitiesnet.com/fn/NS/NS3n9kd.html>

Sullivan, E. (1998a). 'The Whole Building Story: A special report on Energy Star buildings and related strategies for improving energy efficiency'. *Building Operating Management*, September. <http://www.facilitiesnet.com/fn/NS/NS3b8ij.html>

Sullivan, E. (1998b). 'Is the Future Now?: Integrated and interoperable systems are the wave of the future'. *Building Operating Management*, April. <http://www.facilitiesnet.com/fn/NS/NS3b8db.html>

Tatum, R. (1998). 'Sorting Through Integration Options: When almost anything is possible, the real question is, what's worth doing?'. *Building Operating Management*, August. <http://www.facilitiesnet.com/fn/NS/NS3b8hd.html>

Tatum, R. (1999a). 'Energy Upgrades: First Things First'. *Building Operating Management*, April. <http://www.facilitiesnet.com/fn/NS/NS3b9da.html>

Tatum, R. (1999b). 'Interoperability from A to Z: A well-designed building skin can improve occupant comfort while reducing energy use'. *Building Operating Management*, February. <http://www.facilitiesnet.com/fn/NS/NS3b9bh.html>

Urdinola, R. (1996). 'Gamet of Controls'. *Lighting Technology*. <http://www.facilitiesnet.com/fn/NS/NS3b56a.html>

Vergetis-Lundin, B. L. (1998). 'Energy Information Infrastructure: Energy management systems and energy accounting software are key building blocks'. *Building Operating Management*, December. <http://www.facilitiesnet.com/fn/NS/NS3b8lc.html>

Vergetis-Lundin, B. L. (1999). 'Information Management Bandwagon: Energy information management systems have been around for years, but less than 5 percent of all commercial buildings use them'. *Energy Decisions*, March.
<http://www.facilitiesnet.com/fn/NS/NS3n9cb.html>

9

CAD databases

9.1 Introduction

Information has proliferated at an enormous rate, and will continue to do so, causing a gap in the ability of many managers to deal with its mass and variety. The future belongs to the most effective manager of an increasing flow of information, and this will require a well-disciplined use of technology.

As information is an important tool in business, success depends on having the right information in the right person's hands at the right time. Facility managers require tools to enable a proactive management process and that feed into the strategic planning of the organization. The tools used by facility managers should provide valuable information on space, maintenance and capital assets, which can then be used by administrators for their business planning processes.

Information handling is at the core of most essential maintenance and engineering functions; for example, improved customer service, quality enhancement, upgrading safety programs, regulatory compliance, process re-engineering and predictive maintenance techniques. Successful management systems must either highlight what is important or store everything and have good indexing capability so managers can find what they need when they need it.

Many organizations need to be able to access data quickly which may be shared between applications or sites. In response, facility managers must take advantage of the connectivity potential of networks, bridges and gateways using appropriate hardware, application software and cabling. A bridge interconnects networks or systems with similar architectures; a gateway connects networks or systems of different architectures (Kimmel, 1994).

9.2 Computer-aided facility management

Many organizations use software programs that track everything from corporate budgets to furniture and office space. At the forefront is computer-aided facility management (CAFM) software that can integrate drawings, database functions and spreadsheet reporting functions in a graphical layout format. Sophisticated CAFM programs are capable of providing pre-building designs or pre-office designs, strategic planning, construction, move implementation, current facility inventory, budget administration and block planning.

CAFM systems can now interface with each other and can be used for gathering and maintaining accurate facility asset information and establishing and optimizing space within specific office areas. CAFM and computer-integrated facility management (CIFM) systems are characterized by multiple linking applications (for example, space/asset management with work management), either on a single dedicated PC or on multiple, disbursed PCs, using client/server databases that can support a multi-tiered (disbursed server) PC architecture (Teicholz, 1999).

A CAFM survey (Anonymous, 1997) revealed that space inventory/management, followed by space forecasting, fixed asset management, furniture inventory/specification, and maintenance/operations management, comprise the leading applications. CAFM is often used to maintain databases of building outlines (from exterior and core to rooms and spaces), group/departmental or business unit outlines, and furniture, fixture and equipment (FF&E) databases. Interestingly, the performance of software products was regarded as least satisfactory in space forecasting, followed by facility budgeting/cost accounting, technical document management and telecommunications/cable management. At the high end of the scale were CAFM benefits associated with the amount and quality of data available to facility managers. It was found that technology had a clear and positive impact on the productivity associated with the performance of the facility manager's job.

Computer-aided design (CAD) is usually an essential part of any CAFM/CIFM system. CAD enables the graphical representation of buildings and contents, measurement functions and the assignment of 'attributes' that add intelligence to drawn objects. CAD may be either two-dimensional or three-dimensional, although the latter is the most useful as it works on the basis of a single building model. The CAD model is the central database that other applications can access and process.

9.3 CAD systems

CAD refers to a class of programs that assist in modelling and documenting the geometry of a design. CAD drawings and the representation of the information in a CAD drawing differs from a digital image because it explicitly represents the vertices, lines, wire frames, solids and/or surfaces of the object(s) in the design (Anonymous, 1998b). Three-dimensional CAD/CAM/CAE (computer-aided design, modelling and engineering) systems are used to model, evaluate and refine a complex system entirely on the computer, before time and money are spent on the real thing.

CAD software consists of several programs and sub-programs located in a system-specific file structure and often consists of both:

1. General and basic parts (such as elementary drawing routines) which often can be bought off-the-shelf.
2. Application area specific features (such as programmed features, symbol libraries) which form the core of the application.

The user interface refers to the tools and methods the user can employ in communicating and controlling a CAD system, and includes both hardware and software features. Essential components of CAD hardware are a large high quality colour monitor, a keyboard to input exact co-ordinate and numerical information, a mouse and an output device such as a laser printer (A4 and also A3), pen plotter or an ink jet or colour laser for 3D modelling and visualization (Penttila, 1997). Typical features of a CAD software interface are the numerous windows, sub-windows and menus on the screen with which the user gives commands via a mouse or keyboard. CAD application interfaces can often be customized to meet the user's needs.

The working area in CAD applications, often called modelling space, functions with a co-ordinate system comprised of x-y axes in 2D, and a z axis to define height in 3D. Typically there exists several co-ordinate systems: absolute modelling space co-ordinates, window co-ordinates, and users can also define their own 'local' co-ordinate systems. A CAD model is constructed using elementary drawing and modelling elements called basic modelling primitives. Typical elementary CAD primitives are:

1. 2D drawing: points, lines, polylines, circles, arcs, etc.
2. 3D geometric modelling: 3D lines, volumetric boxes, spheres, ellipsoids, torus, pyramids, wedges, cylinders, cones, etc.
3. Object-oriented CAD: 'real' design entities, such as building walls, windows and slabs, etc. (Penttila, 1997).

Line-type and colour are typical visual properties that determine how the primitives appear on the screen and in the output. Parametric representation of basic primitives provides unlimited scaling capabilities. The choice of scale is determined by the level on which particular details are investigated. In 3D CAD the basic primitive elements represented by solid shapes can be joined, intersected and subtracted to generate an unlimited number of shapes, which can be further altered using geometric transformations such as scaling, rotation, translation and stretching. CAD systems also incorporate built-in geometric constraints, for example relations of intersection and bisection, orthogonal and tangent relations, curvature and convexity in order to construct lines and surfaces, and to combine them into realistic objects (Anonymous, 1998b). CAD symbols are prefabricated components created from elementary entities. A symbol is a group that can be moved and deleted as a single primitive.

Information in CAD systems is stored in data folders called layers. Each layer holds specific information, such as the location of exterior and interior walls, or a drawing title block. The amount and naming of layers depends on the user. When drawings are produced by the CAD system, individual layers are combined to provide the needed information. This allows different users to share the same document, whereby each use appropriate layers to create individual documents. For example, one layer may contain data on the size and location of the interior structural supports for an area, another layer may contain data on load-bearing walls; a third may hold information on architectural walls. By combining layers, a person using the system can produce a drawing showing the layout of a space (Piper, 1999).

An attribute is alphanumeric information (non-graphical data) connected to graphic primitives that contains essential design information, for example finishes, materials, estimates of energy consumption and expenses. In a sense, this information becomes an additional layer that is used for generating 3D visual models from the CAD drawings. Attributes can be defined by the user, or they can be derived automatically from graphics or from a database connected to the graphical CAD system, or they can be calculated from graphical information. The use of attribute data connected drawings is referred to as intelligent or integrated CAD which has advanced capabilities such as creating a design model, and the flexibility to exchange CAD data between several projects and with other users.

The association of non-geometric data for CAD objects has implications for all stages of a design, from a more complete documentation of conceptual design to include the designer's intentions to a more complete representation of the final product for facility management. The capacity to handle non-graphical information enables it to maintain all design information in one medium. In facility management, this means that CAD is not only a designer's tool, but also a building end user and owner information database, and can be the main tool in gathering and maintaining all building facility data (Penttila, 1997; Anonymous, 1998b).

Reference files or drawings are used to link CAD files together to form logical structures to resemble reality. CAD drawings made up of combinations of reference files are updated automatically when changes to the original are made. The use of reference files is efficient, since it reduces the need to copy drawing parts into other drawings and it can make the active working files smaller by locating some of the visible information in the reference files. The reference files can not be altered via the visible links, only by opening the original files (Penttila, 1997).

The objective of 3D modelling is usually to describe, analyse or visualize three-dimensional space. A 3D model is sometimes created from 2D existing drawing files, which are completed and detailed more to form a 3D model. Since a 3D model is viewed through 2D screens, the models are represented as axonometric or perspective projections. Digital 3D modelling includes two main steps: (1) generating a wireframe model, in which vertices and edges of an object are explicitly represented and displayed, and (2) rendering that model (Anonymous, 1998b).

An almost unlimited number of different axonometric and perspective views can be obtained by varying the values of the point of view and rotation of axis. Perspective rendering is a technique for constructing 3D realistic visual simulation from a particular viewpoint within or around the model. CAD allows an easy shift between perspective images of 3D compositions and axonometric views, and 2D plans, elevations and sections. The database layer provides the connectivity for different CAD representations. For instance, the same wall can be visualized by projected lines in a floor plan, by another symbol in a building view or by a textured surface in a model.

Static 3D models are the data source for creating walk-throughs, allowing operators to explore design solutions both from 'outside' and 'inside' a model. Such a device allows the designer to simulate the use of their design configurations (Anonymous, 1998a). 3D modelling applications and visualization software also allow the user to create animations and 'virtual reality environments' from 3D models. This requires special hardware and 3D specialized equipment which represents the high end of 3D CAD (Penttila, 1997). Like any computer simulation, the key benefit of virtual designing is that it can be used to design, evaluate and optimize the layout or operation of a facility before it is ever physically built. Simulations are faster and cheaper to build than pilots and changes can be evaluated without disturbing existing operations.

An integrated CAD/CAM/CAE system can allow an organization to make a change in one place and understand all of the ramifications, and it can also become a platform for scheduling and optimizing a facility that is already in operation. Specialized CAD software extends its reach into facility management, and is used to store, retrieve and manage information about a facility and its equipment. Facility managers can modify the data once the facilities are built, making it easier to retrieve it for maintenance purposes and layout changes (Mills and Schmitz, 1998).

Some CAD software enable 3D modelling not drawn from 2D drawings but created as true 3D objects. These systems represent the cutting edge for CAD development, and in particular its application to the facility management discipline.

9.4 Information management tools

A simple inspection of the office will help managers identify the major categories of information that must be filed, retained, accessed and distributed. For example, this may include equipment or asset manuals, parts lists, specification sheets, drawings, field modifications, wiring diagrams, field notes, installation notes and contractor invoices. The first factor to consider is the nature of the information: its use; who uses it and why, how often and for what use; and if it will become obsolete. A careful distinction between information that is held for primarily archive purposes, operational purposes and legal or statutory requirements should be made. Determining what to discard or to archive will free up space and make it easier to find critical information and, as a result, increase productivity (Levitt, 1997).

Various excellent data management tools exist that make life easier for the facility manager, such as those for forming human and corporate contacts with personal information managers, custom designing information systems, and for using the Internet to get information directly from equipment vendors. For recurrent projects, building a database of contacts, organizing the construction files, locating all equipment manuals and setting up a maintenance technical library may be desirable.

There are software packages that assist in a facility audit and strategic space management which aid in tracking, analysing and maximizing depreciation. They produce information that is fully integrated into the financial accounting systems in large organizations, with an automated indirect cost recovery calculation (Henderson-Paradis, 1996). Space management software also can analyse 'what-if' scenarios for planning, delineate depreciation by building, improvements, components, and so on, and outline true allocation of space use costs. Once key space information is gained, targets are set, performance can be measured quantitatively and objectively, and the administrator can make informed decisions for the future.

A problem that many facility managers may face is the lack of resources for adequate upgrades and IT support and consequently they cannot take full advantage of the latest software developments available. An alternative approach is to contract IT services through the Internet and use their software and hardware for a service fee. Any upgrades or new software requirements are absorbed by the IT service company (Piper, 2000).

9.5 Software functions

Capital asset planning is used by the facility manager to chart a long-term plan for facility use and investment, including program changes, physical changes (programmed maintenance), projected capital outlays over a multi-year period, and phasing of cash flows. A capital asset plan ensures that facility investments are prudently made. A capital asset plan utilizes four key tools: market assessment, programmatic assessment, physical assessment and financial assessment. The process of arriving at a capital asset plan combines these four tools and other space management tools to develop viable options and to ultimately arrive at an optimal strategy for long-term facility renewal and regular programmed maintenance (Henderson-Paradis, 1996).

The range of CAFM/CIFM functions is considerable. However, the wide variety of information sources, processes and conditions makes it nearly impossible for a single software package to provide all functions for all facilities. Typically, a given package has two to four significant capabilities. A facility manager may need to select one of these packages and supplement it with another that supplies the remaining necessary functions. A properly selected set of software tools speeds and enhances the facility management process, and results in more efficacious designs, broader consensus and greater competitive capability. When selecting CAFM/CIFM tools, organizations should:

1. Plan ahead with written documents and mission statements.
2. Develop specific, measurable goals (for example, reduce occupancy costs, improve return on assets).
3. Create written procedures on how technology should be used.
4. Set priorities for applications, software and hardware.
5. Have a bias for action. This seems to contradict point (1), but action steps must be part of the plan.
6. Have a financial focus. Try to rank activities in terms of the financial implications.
7. Make a 'business case' for all major activities. Most vendors will help facility managers develop a business case for technology.
8. Perform a simple cost/benefit analysis.
9. Conduct a 'sanity/reality check'.
10. Have a people-focus in technology management (the priorities should be money, people and technology).
11. Avoid the trap of developing customized in-house software. The cost of maintenance is too high with in-house systems.
12. Minimize sources/use fewer vendors.
13. Have a healthy respect for, and fear of, data entry costs (Anonymous, 1998a).

9.6 Buying CAFM/CIFM software

Software buyers should consider the following three questions:

1. Where do I need the most help?
2. How difficult is the software to set up, learn and use?
3. How useful are the results? (Lee, 1998)

Facility management takes place at various levels of detail and requires a structured process to separate these levels into distinct projects or subprojects to limit the problem to a manageable size and keep detail from overwhelming the problem. For each level and task group, different software functions and features are useful. Data acquisition and analysis usually are the most time-consuming tasks, while strategy is the most difficult.

When buying CAFM/CIFM software, it is best to start with a list of everything the organization would like to do with the new system. Obtaining input from employees on how the program could be used, and gaining support from the organization's information management division are also important. The information group can advise on special data that might be needed on networks and languages and whether there are databases available in the organization that can be shared. When selecting potential software suppliers, create a package of material that reflects everything needed to maintain and manipulate the system. If possible, create a review team that includes users of the proposed system to visit vendors and manipulate the company's data on their hardware and software (Anonymous, 1993).

As with any purchase, it is best to shop carefully. Read product reviews and visit trade shows that bring together many system vendors. But be aware that product demonstrations are designed to do exactly what the vendor wants in the best possible manner – they rarely reflect day-to-day use. When properly planned, implemented and maintained, integrated computer systems can give facility managers the information they need to make a real contribution to the corporate mission, according to a recent independent study (Hawkins, 1998).

Implementation of an automated facility management system requires a plan that is supported and advocated by the finance department and understood and sponsored by those in information technology. Facility managers who are tied into the organization's corporate mission have an easier time demonstrating the need for an automated system, particularly if they emphasize maximizing asset value, not just containing costs. Facility managers must become aware of the value of their information and systems to the financial objectives of their organization. Facility managers can use an integrated computer system to help provide the infrastructure and information required to run the company.

Once a facility manager has the corporate and financial approval for an automated system, it is critical to make sure the system runs. After determining the organization's requirements and designing the system, it can be easy to overlook the need to create a data-entry plan and account for its cost. Data collection can use up much of the budget – taking into account routine updating as well as the initial data entry. Accurate data entry is critical to an effective automated system; simplifying the data-entry system is the best way to ensure accuracy and reduce costs.

9.7 The future?

The current generation of facility management software is defined by the use of the World Wide Web/Internet. Current vendor capabilities relate mostly to live (real time) or passive reporting, which is relatively simple to accomplish since all databases include HTML as an output format. Most vendors enable users to execute queries on the Internet (e.g. to display a floor or occupancy plan, to dynamically search for a person or a room via an intranet or extranet) or collect information such as work-request data for a maintenance application. Internet-based applications for managing construction projects are simplifying the process of exchanging information. Data transfer is high-speed and secure, and as data access is in real-time, managers can work with up-to-date information (Piper, 2000).

The next generation of technology will be characterized by widespread chip integration into assets that are tied to the Internet. Thus, for example, a help desk might be notified automatically when there is an equipment or component failure. Notification might include a list of tools to bring and a diagnosis of the failure obtained by the component and the sending of data over the web to the manufacturer for analysis. Likewise, vendors automatically will receive utilization data over the Net on their products, and assets will be tracked by sensors. Web-based energy management systems are available that monitor energy use, while allowing the checking of utility rate structures (Tatum, 2001). Vendors will customize products to specific installations and needs, and will be able to respond rapidly to new requirements and problems. Having embedded sensors in assets will mean that up-to-date asset location, tracking and database updates will be possible. If privacy problems are resolved, embedded sensors and geographical position systems will track individuals, making it possible to automatically configure infrastructures (for example, PC hard drives, telephones, data access) as employees meet in various locations or track time-on-task data. Although there are many positive and negative implications of such pervasive (and invasive) applications of technology, such functionality will most certainly be available.

Probably within about five years the world of objects and object-oriented programming (OOP) and inter-operability finally will have penetrated facility management. Accepted standards of 'objects' will be the defining factor of this sixth generation of facility management technology. In this world, data will be passed between applications in a totally transparent manner, most certainly in multiple multi-media formats, and on some future embodiment of the World Wide Web.

References and bibliography

Alexander, K. (1996). *Facilities Management: Theory and Practice*. E. & F.N. Spon.

Anonymous (1993). 'Facility Assets: A Company's Hidden Cash'. *Office*, 118(4), October, pp. 36-37.

Anonymous (1997). 'Space Functions Lead in '96 CAFM Survey'. *Facilities Design and Management*, 16(2), February, pp. 44-49.

Anonymous (1998a). 'The 13 Commandments'. *Facilities Design and Management*, 17(3), March, p. 51.

Anonymous (1998b). 'Technology and the Virtual Design Studio: How did we get here?', April. <http://www.arch.su.edu.au/~mary/vdsbook/Chapter2/index.html>

Atkin, B. and Brooks, A. (2000). *Total Facilities Management*. Blackwell Science.

Barrett, P. (1995). *Facilities Management: Towards Best Practice*. Blackwell Science.

Cotts, D. G. (1999). *The Facility Management Handbook* (2nd edition). AMACOM.

Hamer, J. M. (1988). *Facility Management Systems*. Van Nostrand Reinhold.

Hawkins, B. L. (1998). 'Strike up the Integrated Systems Band'. *Facilities Design and Management*, 17(7), July, pp. 62-65.

Henderson-Paradis, R. (1996). 'The Strategic Planning Process and the CEO: Facilities managers need the tools, technology and new processes to provide top management with important information'. *AFE Facilities Engineering Journal*.
<http://www.facilitiesnet.com/fn/NS/NS3a96e.html>

Hesselbein, F., Goldsmith, M. and Beckhard, R. (1997). *The Organization of the Future*. Jossey-Bass.

Kimmel, P. S. (1994). 'The Vicious Cycle and New Ways to Manage'. *Facilities Design and Management*, 13(2), February, p. 29.

Lee, Q. (1998). 'Points to Consider in Selecting Facilities Planning Software'. *IIE Solutions*, 30(1), January, pp. 42-43.

Levitt, J. (1997). 'Managing in the Information Age: Technology expands its role as departments seek to streamline operations and maximize resources'. *Maintenance Solutions*.
<http://www.facilitiesnet.com/fn/NS/NS3m7ji.html>

McGregor, W. and Then, D. (1999). *Facilities Management and the Business of Space*. Arnold.

Mills, R. and Schmitz, B. (1998). 'Manufacturing Goes Virtual'. *Computer-Aided Engineering*, 17(12), December, pp. 30-37.

Muller, N. J. (1996). *Network Planning, Procurement and Management*. McGraw-Hill.

Pacanowsky, M. (1995). 'Team Tools for Wicked Problems'. *Organizational Dynamics*, 23, pp. 36-51.

Penttila, H. (1997). 'Lecture Material for a CAD-course, Royal Institute of Technology, Stockholm'.
<http://www.tut.fi/units/arc/aml/CADlect97/CAD2.html>

Piper, J. A. (1999). 'A New Spin on Facility Management: High-powered but underused, facility information technology can put a whole world of data in your hands'. *Building Operating Management*, July.
<http://www.facilitiesnet.com/fn/NS/NS3b9gc.html>

Piper, J. A. (2000). 'Information Advantage: The reach of the Internet is expanding the power of facility information technology'. *Building Operating Management*, April.
<http://www.facilitiesnet.com/fn/NS/NS3b0da.html>

Price, S. (1997). 'Facilities Planning: A Perspective for the Information Age'. *IIE Solutions*, August, pp. 20-22.

Rondeau, E. P., Brown, R. K. and Lapides, P. D. (1995). *Facility Management*. John Wiley & Sons.

Rosenblatt, B. (1995). *New Changes in the Office Work Environment: Toward Integrating Architecture, OD, and Information Systems Paradigms*. Ablex Publishing Corporation.

Tatum, R. (2001). 'Essentials for Internet Energy Management'. *Building Operating Management*, March.
<http://www.facilitiesnet.com/fn/NS/NS3b1cg.html>

Teicholz, E. (1999). 'FM technology: Generation Next'. *Facilities Design and Management*, 18(1), January, p. 18.

Tompkins, J. A. (1996). *Facilities Planning* (2nd edition). John Wiley & Sons.

Risk management

Given an uncertain future, organizations face continual change to remain competitive and to deal with potential impacts over which they have little control. This gives rise to risk exposure that must be appropriately managed. Risk is interpreted differently according to the attitude of the decision-maker, although high levels of risk usually must carry a strong possibility of large financial gains or other benefits to be a worthwhile proposition. While uncertainty is ever present, it does not necessarily translate into high risk exposure, as a decision to move ahead with a particular investment may still represent the best option and deliver an acceptable return.

The difference between uncertainty and risk is at the heart of understanding risk management. Anything that involves making forecasts of future events obviously contains some level of uncertainty that must be assessed. There is a range of deterministic and stochastic techniques that may be employed to do this. However, in the end it is a judgement call based on objective information and historical trends. In other words, planning for future events that will increase profit is somewhat of a gamble, despite there being strategies that can treat risk, such as its transfer to other parties.

Risk is a well-known and practised activity in investment decisions, but is equally vital in facility management. Even ignoring the situation of new investment decisions related to facilities, risk management must cover issues such as outsourcing implications, continuity planning and disaster recovery. There are also environmental health and safety risks, as well as the potential for downsizing and budget cutbacks that are generally pervasive. Risk is a key for the successful management of both new projects and existing facilities.

Nevertheless, risk is linked to decisions, even a decision to do nothing (or no decision). Ignoring risks is to invite catastrophe and not to be prepared for it when it strikes. Contingencies are a common method to divert such catastrophes, either by access to additional funding sources or by the formulation of plans in anticipation of problems that might arise.

Decision-makers fall within a spectrum described as having risk-adverse people (conservative) at one end and risk-seeking people (speculators) at the other. While projects or events may be identical, the interpretation of them by different types of decision-makers may lead to inconsistent recommendations. Therefore risk assessment is as much a function of the context of the decision-maker as the nature of the activity itself.

Risk management is normally defined as encompassing risk identification, risk analysis and risk treatment. Identification concerns discovering the areas which have the most significant impact on the expected outcome or which are the most uncertain. Analysis involves determining, in a quantitative and objective way, the level of risk under a range of possible scenarios (often described as best and worst case). Treatment involves developing strategies to limit these risks or to get back on track if the worst case scenario becomes reality.

Many organizations use outsourcing to transfer risks associated with non-core business activities to other organizations that have specialist knowledge or capacity. Although the concept carries risks of poor performance or default, these are dealt with in contractual agreements and the setting of key performance indicators that must be met. Outsourcing can minimize an organization's exposure and the costs of developing appropriate responses to ensure performance expectations are met. Usually such advantages have a cost penalty that must be balanced by increased service, quality or the like to make the proposal worthwhile.

One principal risk for many businesses is the prevention of trade due to some unexpected event. Continuity planning is about ensuring that there are procedures in place to recover from a disaster should it ever occur. These plans are necessarily broad and must deal with potential disasters that arise from internal and external sources. Loss of revenue can be significant in such cases. Insurance is a common strategy for dealing with the cost implications of disasters, but other plans are necessary to ensure that safety is maintained, there is security over assets (including electronic data) and that any reconstruction can occur in minimum time.

This part concerns risk management and its importance to all types of organizations. Decisions involving uncertainty that do not comprise an assessment of risk are likely to lead to poor outcomes unless by chance. Chapter 10 discusses the difference between risk and uncertainty in more detail. Chapter 11 looks at the popular activity of outsourcing; its advantages and disadvantages, in the context of risk treatment. Chapter 12 deals with continuity planning and disaster recovery strategies.

Ignoring risk is a negligible action when dealing with future events. Facility managers need to understand and apply risk management techniques and develop contingencies against unforeseen events. Scenario analysis is useful in this regard and forms the basis of modelling uncertainty in a quantitative and repeatable fashion.

Risk and uncertainty

10.1 Introduction

The forecasting of future events is normally an integral part of the decision-making process. It also can be the subject of considerable uncertainty and therefore requires cognizance of the level of risk exposure. Facility management activities are clearly reliant on appropriate forecasts of future events being made.

The decision-maker is faced with a fundamental choice. Either the uncertainty of the future can be ignored by dealing with only those matters that are known, or tools can be used to help make predictions. It is frequently accepted in the literature that it is preferable to plan even when the accuracy of the plan cannot be guaranteed, otherwise decisions are made in isolation to the environment in which they will ultimately be judged.

Most forecasting techniques will inevitably fail to predict catastrophic events but instead will focus on a range of outcomes that may be reasonably concluded from history. This results in the identification of best, worst and most likely outcomes. If a decision is insensitive to this range then it has a lower level of risk than if it were to change easily. Forecasting adds information to the decision-making process that is a vital part of proper analysis.

The adoption of risk analysis techniques enables the uncertainty of future events to be properly assessed. For example, if the interest rate applied to a new development is likely to vary within a range of values, but at all values the project outcome is favourable, then what is at risk is the level of the benefit, not the possibility of a loss. In this case, uncertainty remains high but risk of financial loss is low. Facility decisions are not always so simple, and therefore a number of sophisticated risk analysis techniques are available to quantify the impact of various future scenarios.

10.2 Forecasting future events

Risk management decisions, particularly in the area of risk treatment measures, often require forecasting future events of accidental losses. Forecasting future events involves studying past events to identify trends, and projecting these trends into the future. The trend or pattern can indicate stability or change, but even for the latter, there could be an element of constancy, such as a known rate of change (for example, inflation or frequency of work injuries proportional to output levels). To forecast future events under the 'no change' trend, probability analysis is an appropriate tool of prediction. Under the circumstances of predictable change, regression (trend) analysis is the best method of forecasting events.

Probability analysis is most effective if a lot of data on consistent past losses is available and the organization's operations have been fairly stable. In this case, the projection of past trends into the future will be fairly uncomplicated and predictions will be reliable, except in the case of unforeseen and uncharacteristic events, such as changes in operations, technology, personnel, physical environment and disasters. The probability analysis will estimate the probability or relative frequency and standard deviation with which an event can be expected to occur in the future in a stable environment. The accuracy will depend on the sample size and how representative it is of the event to be predicted. Regression analysis also projects past trends into the future, but it also factors in any concurrent patterns of change such as, for example, changes in loss frequency together with changes in another, more easily measured variable, such as time. Two or more trends might move together, and knowing one helps forecast the other. The predictable variable, also known as the independent variable, such as time, sales, production, personnel, etc., is related to the dependent variable to be forecast (Head, 1995).

10.3 Levels of uncertainty

Uncertainty refers to the state of knowledge about the variable inputs and outcome of an analysis. Decision-making is easiest from a position in which the future can be predicted with a degree of certainty, but more often the future is far from certain and is subject to the influences of a multitude of unpredictable factors. There is a natural compulsion for managers to act decisively and to offer clear strategies when the future is least clear. Drawing up strategy in uncertain environments is made more difficult by the fact that many of the ideas and concepts about strategy formulation were advanced in an era where the future was probably more predictable (Courtney et al., 1997). Decision-makers need to properly understand as much as possible the uncertainty being faced and make best use of, and possibly influence, an uncertain environment.

Courtney et al. (1997) offer a flexible approach to predict the future and evaluate strategic options that involves identifying the type of unpredictability that exists, selecting a strategy that provides an answer to it, and using it to positive advantage. They describe four levels of uncertainty, each requiring a different approach when making strategic choices:

1. *Level 1: Clear enough future*. In this case the general characteristics of the environment and its dynamics are understood. The information needed to make the analysis for a strategic decision will either be known or can be found out by market research or competitor analysis. This is the most common type of uncertainty level and is amenable to analysis using traditional frameworks.

2. *Level 2: Alternate futures*. In this case the future is defined by discrete scenarios, for example passing of legislation that affects the organization's market, or a competitor that disturbs pricing or supply chain relationships. Although it is difficult to predict precisely what outcome will take place, game theory and option valuation will help to create a clearer understanding of what the chosen strategies for the organization should be.
3. *Level 3: Range of futures*. The future may fall within a describable range of outcomes, but where exactly within that range remains unpredictable, for example whether to enter a market where response to a product or technology is difficult to predict, or whether to invest in emerging technologies. Here, the advice is to develop a limited range of scenarios that describe possible alternative, but meaningful, futures and to use these to assess whether the chosen strategy makes sound financial and strategic sense.
4. *Level 4: True ambiguity*. Uncertainty is very high and the outcomes are affected by numerous variables. Although managers might form a strategy based more on instinct than reason, they need to learn from how similar markets or technologies have developed. If information is insufficient to make the decision to invest, then information needed to take a decision will need to be found.

10.4 Risk identification

The facility manager must be concerned with any circumstance that generates potential losses to the facility, and is usually well placed to identify the critical components of the organization in the role of maintaining steady and efficient operation. Without the recognition of risks, or potential sources of loss, no evaluation of risk or provisions for handling can follow. These risks may include physical injury to occupiers or visitors, or loss of/damage to property, and the risk of injury to third parties or damage to their property, for example through the spread of fire. Risk identification is essentially about answering three simple questions: what can happen? how can it happen? and why can it happen? There is a range of techniques available for undertaking risk identification. Choosing the most effective technique for a particular circumstance will depend on the nature of the circumstance and the desired outcomes, for example financial risks require techniques involving review of balance sheets and cash flows; or risks associated with a building will be best identified by physical inspection. Various sources of information are useful to identify areas of risk:

1. *Experience and records*. This includes insurance claims, workers' compensation claims, incident reports, incidents in similar organizations or within industry in general, historical data relating to natural events, systems failures, fraud, theft, operational losses, fines, penalties, complaints, etc.
2. *Brainstorming among people who know and run the system*. For example, plant engineers, area managers, supervisors and equipment operators.
3. *Physical inspection*. This enables 'first hand' appreciation of activity and permits consultation with managers and employees to assist in identifying risks. Reports from previous physical inspections/audits are also a useful source of information.
4. *Organization charts*. These are useful in illustrating different aspects of the

organization's activities and structure, and can be used to pinpoint areas of risk, for example dependencies, duplications and concentrations (Williams, 1999).

Using flow charts (production, service, accounting, marketing, distribution, etc.) as a risk identification tool highlights the effect of certain events on the overall flow, for example using a service/production flow chart will enable assessment of the impact of the loss of a particular service/item on the organization's production capacity. A checklist can be used to provide an overall picture of the sources of risk for an organization. Future iterations of the risk management system will allow this picture to be modified to suit the particular circumstances. Other risk identification techniques include failure mode and effect analysis, human reliability analysis, event tree analysis, fault tree analysis, HAZOP (hazard and operability), human reliability analysis, preliminary hazard analysis and reliability block diagrams (Williams, 1999).

Comprehensive identification using a well-structured systematic process is critical, because a potential risk not identified at this stage is excluded from further analysis. The most effective risk identification is performed in consultation with the organization's personnel.

10.5 Risk analysis

Risk assessment is comprised of two components. The first component involves evaluating the probability of a risk occurring and the impact it will have when it does occur. The second component involves ranking the risks once the probability and impact are estimated (Stevens, 1997). The objective is to separate minor (acceptable) risks from major (unacceptable) risks, and to provide data to assist in assessing the appropriate treatment of risks. Part of risk analysis involves identifying the existing management and technical systems and procedures in place to control risk.

Sources of information to estimate probability and impact may include past records, relevant experience, specialist and expert judgements, industry practice and experience, published literature, test marketing and market research, experiments and prototypes, and economic, engineering and other models (Williams, 1999).

The risks are assessed in terms of their probability of occurring in the organization. This would range from 'possible but very unlikely', through 'low probability' to 'high probability with an expected frequency greater than once per year'. Predicting the impact of each risk is described in terms of high, medium and low and involves consideration of various areas of impact, for example cost, people, environment, legal liability, public perception, etc., and relies on the systematic use of contextual information. Once the probabilities and impact are evaluated, the risks are then ranked as either a major, moderate or minor risk (for example, a major risk is one with high probability and high impact). There can be many permutations and the ranking will depend on the nature of the project and the level of caution exercised by the decision-makers (Stevens, 1997).

It is important that risk management criteria are established before the risk analysis process begins. These criteria will be strongly influenced by the organization's contextual circumstances. If risks are low, they can be accepted with minimum further treatment, but they should be monitored and periodically reviewed to ensure that they remain acceptable. Risks that do not fall into the low or acceptable category should be treated where the greater the risk, the higher the priority.

10.6 Risk treatment

Risk handling devices, such as assessing exposures and securing suitable insurance policies or other loss-funding arrangements, have been the main precautions taken by organizations to bear the financial impact of a loss or disaster. However, this fiscal approach is often not adequate for facilities of all sizes, and as a result, risk management is being executed with an approach that often focuses more on the control, that is, loss prevention and loss minimization of potential losses, as well as on more creative loss financing alternatives (Barton and Hardigree, 1995). A risk management policy, formed by the collaboration of the facility manager and the senior management of the firm, will assist in making decisions regarding the methods of treatment of loss exposures, levels of retention in the use of insurance policies, etc.

The general techniques of risk treatment are as follows:

1. *Avoid*. Reduce probability of loss to zero (for example, cease activity, close facility, sell business). Be careful not to achieve risk avoidance through an attitude of risk aversion. Risk aversion is characterized by decisions to avoid or ignore risks regardless of the information available, failure to treat risks, leaving critical choices/decisions to others, deferring decisions which cannot be avoided, and selecting an option because it represents lower risk, regardless of the benefits.
2. *Accept and finance*. Risk financing techniques include commercial insurance, contractual transfer, borrowing, current expensing of losses, unfunded reserve, funded reserve and 'captive' insurance.
3. *Reduce likelihood*. Employ risk prevention techniques (for example, compliance programmes, inspection and process controls, security devices, alarms, preventative maintenance, training and education).
4. *Reduce consequence*. Pursue risk reduction (for example, disaster and contingency plans, medical and first aid procedures, off-site data and information storage, fraud control, fire suppression).
5. *Transfer*. This involves another party sharing the risk (for example, contracts, insurance, partnerships, joint ventures). Risk transfer to other parties, or another physical location, does not reduce the overall level of risk to society. Risk transfer introduces new risks (i.e., the other party may not manage the risk effectively, or the transfer mechanism may not prove effective) such as public image, vicarious liability.
6. *Retain*. After risks have been treated, there may be residual risks that are retained. Be careful not to retain risks by default, in other words do not fail to identify or effectively treat risks (Williams, 1999).

Risk treatment options are not necessarily mutually exclusive and may not be appropriate in all circumstances. Risk treatment options should be reviewed according to their feasibility, cost and effectiveness in eliminating or reducing the risk. Selection of the most appropriate option involves balancing its cost against its benefits, and the cost should, in general, be less than the benefits obtained. For effective measurement of costs and benefits, look at the organization's strategic goals, and the performance measures put in place to monitor the progress toward achieving those goals. This reinforces the strong link between risk management and the organization's strategic planning process. As risks rarely remain static, risks and the effectiveness of the control measures need to be continually monitored to ensure the management plan remains relevant.

10.7 Risk attitudes

Risk attitude is the willingness of a decision-maker to take a chance on an action or investment of uncertain outcome. Risk exposure is the probability of investing in a project whose economic outcome is different from what is expected. A given level of risk exposure may be acceptable to a risk taker, but unacceptable to a person or corporation who is risk averse, therefore to make effective choices when outcomes are uncertain, a decision-maker needs to incorporate risk attitude into their evaluations. Decision-makers may be defined as risk-seeking, risk-neutral or risk-averse.

Two general approaches exist to include risk attitude in a project evaluation (Ruegg and Marshall, 1990). A decision-maker can make a decision based on their subjective or intuitive perception of the acceptability of the degree of risk exposure indicated by an analysis of the project's worth. This approach lacks any standard procedure of measuring risk attitude when making a decision, but will be adequate for simple situations such as comparing the predicted outcomes, for example benefit-to-cost ratios, of various options. A more formal approach that uses a measure of the decision-maker's risk attitude to evaluate the worth of a project is based on utility theory. Decision analysis is an example of a technique that uses utility functions to account for risk attitude. It uses decision trees (a recognized risk analysis technique) to represent all possible outcomes, costs and probabilities associated with a given problem. For a description of how to quantify an individual's or organization's risk attitude, refer to Spetzler (1968).

10.8 Managing risk

Risk management can be defined as the practice of awareness of potential risks, the assessment and quantification of them, and the action taken to reduce or remove them where possible (Carmichael and Gartell, 1994). Many of the risks inherent in any building may be minimized or removed by good management. The effective management and control of risk will provide long-term insurance cost reduction, minimize disruption in the event of a loss, and enable a business to operate more effectively.

If resources were infinite, the concerns of each area in an organization could be addressed so as to satisfy each consequence of potential risks. However, most organizations' risk control resources are limited and need to be spread to cover the whole organization. Effective allocation of these resources requires an overall perspective of all the risks and their potential effect on the strategic goals of the organization (Williams, 1999). This holistic approach to risk management heavily relies on line management involvement. Line managers are best placed to control the allocation of resources where they are the most cost-effective against risk. They most fully understand the operations and systems in place, and are able to identify and consider all risks at the same time. The approach has many other benefits:

1. Efficient resource allocation, allowing for prioritization of risk and the cost-effective allocation of resources.
2. The commitment of executive management, which is an opportunity for effective leadership.
3. Unbiased risk/benefit decision-making, where all assumptions, guesses and other intuitive ideas about potential risk are discussed and assessed in the context of operational knowledge at all levels. This allows risks to be offset with benefits in an impartial way, as well as showing the significance of one particular risk within the total range of possible risks. The process effectively sorts out what is urgent, what is important, and what can safely be ignored.

For risk management to be effective it must be considered as an integral part of the strategic planning process of the organization. Ineffective risk management leads either to losses that cripple the organization, or to lost opportunity if the organization is too risk averse. Risk management enhances profitability through the avoidance of loss scenarios that increase costs, decrease revenues or prejudice assets.

Qualifying risk management goals as strategic goals, and considering risk management as part of the strategic planning function within the organization is an effective way of developing a risk management culture within the organization. Executive management objectives for, and commitment to, risk management should be documented in a formal risk management policy. Employees and other key stakeholders (shareholders, community, etc.) should be made aware of this policy. Responsibilities for risk management should be clearly defined. Each stage of the risk management process should be documented. These documents should include assumptions, methods, data sources and results, and should:

1. Provide a record of risks.
2. Provide decision-makers with a plan for approval and implementation.
3. Provide an accountability tool.
4. Facilitate monitoring and review.
5. Provide an audit trail.
6. Enable sharing and communication of information (Williams, 1999).

Employees need to be informed about what the organization's risk management goals are in order to foster a risk management culture. Additionally, they must be kept informed about progress toward meeting these goals to keep them motivated.

10.9 Assessment techniques

The techniques used in risk assessment are varied, including structured interviews, multi-disciplinary groups of experts, individual assessments using questionnaires, computer and other modelling, and use of fault trees and event trees (Williams, 1999). The level of sophistication and degree of accuracy of the risk analysis will depend on the information and data available and the assessment technique used. The analysis may be qualitative, semi-quantitative or quantitative.

Qualitative analysis (for example, event trees, failure analysis, SWOT analysis, checklists and questionnaires, physical inspections), which employ descriptive scales to describe risk, is used as a screening tool where numerical data is unavailable or inadequate, or where the level of risk does not justify the effort required for a fuller analysis. In semi-quantitative analysis, qualitative scales are given values that do not necessarily bear accurate relationship to the magnitude of the risk. Although the values can be manipulated arithmetically, the numbers mean nothing in absolute terms, only relative to other numbers generated in the analysis. Quantitative analysis (for example, computer modelling, fault tree analysis, HAZOP, hazard indices, statistical studies) uses absolute numerical values. The quality of analysis depends upon the accuracy and completeness of the numerical values used. Qualitative analysis, although sometimes regarded as having limited value, can be very useful, particularly when used together with quantitative data.

Quantitative analysis is often more expensive because of the software and expertise required to carry it out, therefore an assessment of its necessity and cost-effectiveness is required to warrant the expenditure (Williams, 1999). The decision on a particular approach should be based on the uncertainty and risk inherent in the capital or investment decision being made. Risk may be defined according to a continuum of levels, and decision-making tools applied accordingly. For example:

1. *Low risk*. The capital or investment decision in question is relatively simple (the purchase of a lorry or a photocopier, for example), and a well-defined forecast can be developed for potential outcomes. Discounted cash flow (DCF) analysis, or even a simple payback analysis, would suffice.
2. *Moderate risk*. Decisions on this level may have multiple outcomes but, generally, potential results are understood with the same level of detail as in low-risk capital expenditures. From a capital budgeting perspective, straightforward DCF can still be applied, but with the addition of sensitivity and risk analysis.
3. *High risk*. Outcomes in this category are hard to define and forecasting involves broad generalizations often centred on 'best-case' and 'worst-case' scenarios. High-risk investments might include acquisitions or the purchase or divestiture of a non-core business. Decision-making for high-risk investments should go beyond financial analysis and emphasize strategic implications. Such complex tools as the Black-Scholes model and Monte-Carlo method used for sensitivity analysis become more important (Williams, 1999).
4. *Extreme risk*. These are all-out decisions (for example, a major shift in strategic direction with entry into new markets and new products) characterized by very little knowledge about outcomes and serious consequences for incorrect choices. Capital budgeting analytical tools are of little use, and risk mitigation and business instinct become critically important. The only tools that really apply are performance measurements to monitor progress, a well-defined and documented exit strategy, and an understanding of how the extreme-risk decision fits into the organization's overall business strategy (Klimas, 1999; Williams, 1999).

Decisions as to the appropriateness of methods may also take into account variations in strategy or market dynamics. Business units that operate in highly volatile markets, where business instinct and market knowledge are very important, require fast decisions. In such cases, an expeditious technique such as the payback method would be more suitable. A business unit that requires a slower, more detailed study such as building a full business case, may require, for example, a sensitivity analysis and a DCF analysis. Units involved in international markets may require special analysis or investigation up to and including the use of outside resources to assist with their decision.

References and bibliography

Alexander, K. (1996). *Facilities Management: Theory and Practice*. E. & F.N. Spon.

Atkin, B. and Brooks, A. (2000). *Total Facilities Management*. Blackwell Science.

Barrett, P. (1995). *Facilities Management: Towards Best Practice*. Blackwell Science.

Barton, L. and Hardigree, D. (1995). 'Risk and Crisis Management in Facilities: Emerging Paradigms in Assessing Critical Incidents'. *Facilities*, 13(9), August, pp. 11-14.

Carmichael, D. and Gartell, S. (1994). 'Insurance and Disaster Planning'. *Facilities*, 12(1), January, pp. 9-11.

Cotts, D. G. (1999). *The Facility Management Handbook* (2nd edition). AMACOM.

Courtney, H., Kirkland, J. and Viguerie, P. (1997). 'Strategy Under Uncertainty'. *Harvard Business Review*, Nov-Dec.

Hamer, J. M. (1988). *Facility Management Systems*. Van Nostrand Reinhold.

Head, G. L. (1995). *Essentials of Risk Control Vol. 1* (3rd edition). Insurance Institute of America, Pennsylvania.

Klimas, A. J. (1999). 'Developing a Value-adding Capital Budgeting Process'. *KPMG Consumer Markets*, August.
<http://usserve.us.kpmg.com/cm/article-archives/actual-articles/1-27-99ak.html>

McGregor, W. and Then, D. (1999). *Facilities Management and the Business of Space*. Arnold.

Moos, R. H. (1996). 'Understanding Environments: The key to improving social processes and program outcomes'. *American Journal of Community Psychology*, 24, pp. 193-201.

Park, A. (1994). *Facilities Management: An Explanation*. Macmillan.

Rondeau, E. P., Brown, R. K. and Lapides, P. D. (1995). *Facility Management*. John Wiley & Sons.

Ruegg, R. and Marshall, H. (1990). *Building Economics: Theory and Practice*. Van Nostrand Reinhold.

Spetzler, C. S. (1968). 'Establishing a Corporate Risk Policy'. In proceedings of the American Institute of Industrial Engineers, Norcross, Georgia.

Stevens, D. (1997). *Strategic Thinking: Success Secrets of Big Business Projects*. McGraw-Hill.

Tompkins, J. A. (1996). *Facilities Planning* (2nd edition). John Wiley & Sons.

Williams, S. (1999). 'Practical Tools and Techniques for Identifying and Analysing Risk'. FMLink.
<http://www.fmlink.com.au/images.au/Papers/nmec.html>

Outsourcing

11.1 Introduction

Outsourcing was originally defined as hiring an outside firm to supply services that were performed in-house. This definition has been broadened to include a way for companies to sharpen their focus. Outsourcing is a management technique to assign peripheral functions to outside organizations (Hidaka, 1999), so that companies can reduce their expenses, focus on their core business and strategic direction, and increase capital and surplus. It is a technique to realign corporations around core competencies and long-term outside relationships and is best driven by clear tactical and strategic goals. Facility management accounts for 15% of all outsourced corporate services in the US. In 1998 85% of organizations were buying services once performed in-house, and this is expected to further increase (Corbett, 1998).

Outsourcing is essentially a risk treatment strategy. While outsourcing may come at a higher price than in-house provision, productivity improvements and economies of scale can deliver overall price benefits, plus open up a number of strategic advantages to the organization. The risk of poor performance is transferred to an external organization better capable of handling the risk and more likely to succeed. In other cases, the price of in-house services is higher due to the lack of competition and over-provision.

There is a clear difference between outsourcing and the engagement of consultants. The latter buys expertise to advise and assist with largely in-house operations, whereas the former is more akin to farming out areas of non-core activity which are better handled by external organizations.

11.2 Reasons for outsourcing

An outsourcer can provide specialized expertise which a company cannot justify developing in-house. Because they have a highly efficient and flexible operation already established, outsourcers are under cost and deadline pressures to perform at a higher level than in-house departments. In addition, outsourcers make each project a priority, not just another good idea that is sidetracked by the pressure of day-to-day operations. Outsourcers are also able to provide objectivity that in-house staff members may be unable to offer (Salvetti and Schell, 1995).

According to a 1998 survey by Jericho, a New York-based Outsourcing Institute, the top five tactical reasons why companies outsource are (Corbett, 1998; Raiford, 1999):

1. *To reduce and control operating costs.* This is done by leveraging the unique expertise and economies of scale and scope offered by the provider.
2. *To make capital funds available.* Outsourcing reduces the need to invest capital funds in non-core business functions, making them more available for core areas.
3. *Cash infusion.* Equipment, facilities, vehicles and licenses used in the current operations all have a value and are, in effect, sold to the provider as part of the transaction resulting in a cash payment.
4. *Resources not available internally.* Companies outsource because they simply do not have access to what is needed. This can be particularly important when expanding operations.
5. *Function difficult to manage or out of control.* Some outsourcing decisions are made simply because an organization cannot do the job well in-house.

Strategic reasons for outsourcing are:

1. *Improved business focus.* Outsourcing operational details to an outside expert lets the company focus on broader business issues.
2. *Access to world-class capabilities.* By the very nature of their specialization, outsourcing providers bring world-class, often worldwide, resources to meet the needs of their customers.
3. *Acceleration of re-engineering benefits.* Outsourcing enables an organization to immediately realize the anticipated benefits of re-engineering by having an outside organization that is already re-engineered to world-class standards take over the activity.
4. *Shared risks.* When companies outsource, investments are made by the service provider who is better able to weigh the alternatives and spread the risks across multiple clients.
5. *Freeing resources for more strategic activities.* Outsourcing permits an organization to redirect its resources from non-core activities toward activities that have a greater return in serving the customer.
6. *Forces outside the organization.* External forces also drive outsourcing, such as deregulation in the telecommunications industry.

The benefits are numerous, but some organizations still hesitate to outsource services for reasons such as fearing loss of control, or questioning the loyalty of workers not directly employed by the company. These are valid concerns, but the key to successful outsourcing is to make the benefits work to your advantage. By outsourcing, a higher level of expertise can be gained at an overall lower cost. Outside experts can not only help re-engineer company processes to make them more efficient, but they can also apply the most useful new technology and ultimately raise productivity to reduce costs. The estimated overall savings through outsourcing is between 20% and 40% (Jurney, 1995).

Successful outsourcing allows companies to redirect their resources from non-core activities to primary activities, thus maximizing performance and allowing them to transform how they do business. Increasingly, companies are choosing service providers that will partner as a vital aspect of their corporation's strategic vision. Starting in the late 1980s as a solution to cost pressures, partnering with outside firms in information technology and other cost-intensive fields proved to be a way to make significant savings, especially major equipment expenditures. A shift from purely cost considerations to business strategies occurred in the 1990s and the focus has grown to include improved expertise and competitive edge as well as cost considerations (Raiford, 1999).

11.3 Most common activities outsourced

The most common activities outsourced by more than 90% of organizations include such things as housekeeping, custodial and grounds maintenance. Other basic activities routinely outsourced include food services, copy centres, mailroom operations, payroll, benefits administration and training. On the increase are the more highly integrated activities, such as logistics, information technology, finance, and marketing and sales (Corbett, 1998). According to the Outsourcing Institute, each year more and more corporations are utilizing outsourcing to introduce new technologies and innovation to their companies. This is especially true in information technology where the client gains a competitive advantage with cutting-edge software and computer equipment, as well as talented technicians and technical support. There is also an increasing trend to outsource intellectual service activities, such as legal services, as well as physical properties (Raiford, 1999). Marketing research, logistics, accounting, legal work and product development require more specialized co-ordination than traditionally outsourced services. Outsourcing growth rates are increasing dramatically and the Outsourcing Research Council, Poughkeepsie, New York, predicts the greatest growth will be in information technology (Raiford, 1999).

Organizations are also selecting providers to re-engineer their departments to increase overall productivity and functionality. In the past, companies often re-engineered departments internally first, then negotiated from a position of strength with outsourcing firms. Now, by working with business partners during the re-engineering process, organizations are able to take advantage of advice from leading consultants and a provider's experience. By relying on outsourcers' expertise while restructuring departments, companies can save money and facilitate the transition period (Raiford, 1999).

11.4 Facility management

Outsourcing redirects the future of business for facility professionals. In addition to cost savings, partnering with outsourcing vendors is being used in novel ways to strengthen corporate missions. To reap the fullest potential from outsourcing and create mutually beneficial outsourcing relationships, facility managers must develop good skills in business management, leadership, negotiation and communication (Raiford, 1999).

Facility management outsourcing extends beyond traditional activities, such as maintenance, landscaping and security, to include diverse functions such as human resources, website technology and all aspects of facility management responsible for the sophisticated systems that are an integral part of today's buildings. In some cases, facility management has been combined with overall property portfolio management.

In order to ascertain how facility managers can contribute to the corporate mission, they should consider a series of questions, such as:

1. What, if any, aspects of facility management represent core competencies of the organization?
2. How might outsourcing of those non-core areas contribute to driving shareholder value?
3. How does the facility manager create unique value for the organization?
4. How do new office approaches change the role of the facility manager and how can outsourcing be part of that solution?
5. How might innovative relationships, like some of those described above, drive greater value? (Corbett, 1998).

11.5 Outsourcing management

Treating outsourcing as a means to relinquish responsibility and leaving key functions to be managed entirely (and possibly improperly) by the service provider will most likely lead to disastrous results. A higher level of involvement and use of management skills are required to make a success of outsourcing arrangements. Important key elements are regular and ongoing communications through formal and informal channels, integrity, innovation and personal service – like any other management situation (Raiford, 1999). Working with partner businesses requires lateral leadership with good skills in negotiation and problem solving. Relationships should be structured in a way that creates a true partnership between the provider and the user with clearly defined, common goals. Acceptable performance levels and exactly what is required should be decided in advance – this should include the response to less-than-acceptable performance levels from service providers. Successfully managing across boundaries leads to mutually beneficial outsourcing relationships.

Outsourcing demands forethought and unique management skills. The initial step when embarking on partnering with service providers is to truly understand the operation that is being outsourced. Facility managers should consider if the task could be properly handled by an external company. Some tasks are so intricately bundled with other business functions that related tasks may need to be outsourced as well (Raiford, 1999).

To decide if outsourcing is appropriate for the company and for the facility department, first consider several basic questions:

1. Do you have, and can you maintain, the staff expertise capable of providing the level of support needed to respond to both internal and market requirements in a timely and effective manner?
2. What is the price of having staff perform these functions and is this the most valuable use of their talents?
3. Can you afford the investment needed to keep pace with the rapid evolution of technical and related expertise for handling more demanding and changing processing requirements? (Salvetti and Schell, 1995).

If the organization is not equipped to handle all of these challenges, outsourcing can provide a viable solution and will eventually save management both time and money.

Organizations go through a definable process as they evaluate, plan, implement and manage outsourcing relationships. The six steps of the process are as follows (Corbett, 1998; Jurney, 1995; Raiford, 1999):

1. *Strategic analysis*. Companies are most successful when they view outsourcing fundamentally as a tool for organizational change. They focus on core competencies both of their own and those of potential providers. Successful organizations also clarify their organizational goals using tactical and strategic reasons that companies outsource.
2. *Identifying best candidates*. What areas are the best candidates for outsourcing? Within the organization, what are the non-core areas where the best return on the investment from an outsourcing decision will be realized? Return on investment is one of the most critical questions to be answered in any outsourcing evaluation.
3. *Defining requirements*. The organization has to define its unique requirements in clear, complete and measurable terms. Because outsourcing is a service, organizations spend a great deal of time describing the services they desire and the results they expect to be achieved via the contract.
4. *Selecting provider(s)*. In terms of selecting partners, customers are increasingly looking for cultural fit. They are selecting providers with whom they recognize a similar way of approaching problems, a similar set of values, and similar criteria in terms of the way they manage their businesses. According to the Outsourcing Institute, the leading factors in vendor selection are:
   - References/reputation
   - Flexible contract terms
   - Commitment to quality
   - Scope of resources
   - Additional value-added capability
   - Cultural match
   - Existing relationships
   - Location
   - Price.
5. *Maintaining control*. Choosing outsourcing does not mean surrendering control over operations. If anything, outsourcing gives executives even greater control through an alliance effort with the service provider. The goal of the service provider is to deliver a higher level of service than customers could ever hope to achieve on their own. Therefore, a facility manager should seek an outsourcing

provider that offers the most appropriate processes, people and technology to meet the company's needs. A provider must demonstrate extensive knowledge of current and emerging technologies, and how to integrate them for maximum effectiveness. Existing equipment must be evaluated for its effectiveness.

6. *Transitioning the operation*. Decision-makers should find a provider who can offer a custom-tailored solution that creates no interruption in workflow. Human resource issues are critical during the transition since employees will be offered other positions within their current company, positions with the new service provider, or some type of severance package for leaving the firm. Most companies now recognize the importance of treating human resources as a line function throughout the process. If the provider hires and retains some of the company's employees, it can have the effect of increasing their productivity while acquiring valued insights about the organization.
7. *Managing the relationship*. Finally, organizations successfully using outsourcing as a management tool have recognized that they must put as much time and energy into managing the relationship as they put into defining the relationship.

A cohesive management process that can be duplicated and repeated is important for both customers and their service providers, and gives a means for facility managers to create value for their organizations.

11.6 Ensuring success

If you decide that outsourcing is the correct alternative, how do you make sure it works for you? In the words of Tom Peters, 'Do what you do best, outsource the rest.' Take a very broad approach and ask yourself, 'What business are we in? What business are we not in?' This basic evaluation of your business mission will help you bring clarity to outsourcing issues (Salvetti and Schell, 1995).

Take time and shop around, and meet with several service providers to get a better idea of the kind of talent that is available. Choose a company with a high level of expertise and check its references. The time invested in the initial search will reap a better long-term business relationship. Also, sign up for a finite trial period of not more than one year. After the first year, an evaluation of the investment should be made:

1. Have profits gone up?
2. Has an advantage been gained in the marketplace?
3. Were the services provided at a higher skill level than those available internally?

A thorough review of the success of the venture into outsourcing will increase the awareness of how additional outsourcing activities can strengthen the business and position it for even more success in the future.

11.7 Strategic alliances/partnering

Some clients and vendors are creating stronger and more long-term, comprehensive relationships called strategic partnering or alliances, based on trust and commitment. Similar to the philosophy that drives outsourcing, strategic alliances mean focusing and capitalizing on what various organizations are actually good at. As these partnerships are intended to last, they provide the incentive for organizations to actually expand and put into practice additional capabilities and offer more comprehensive services.

Companies expanding their core competencies to better serve valued clients are getting in return a corresponding expanded range of commitments from their clients. For example, for a firm that supplies a particular product, additional services could include global availability, installation of its products, workplace consulting services and dedicated account managers. In exchange for the additional services, clients may deliver a number of promises, the nature of which depends on the partnering relationship, for example a commitment to use them as their global or regional provider; agreeing to an exchange of services or promising to use the company as a consultant (Sommerhoff, 1998).

This trend has been stimulated by the current nature of business – global, fast and highly competitive – which has necessitated streamlined, efficient services. The relationships are usually developed over a considerable period of time, but once established, they are founded on trust and commitment and, of course, a mutually viable financial contract. They also require dedication from both sides.

To assuage any fears about how loyal and dedicated clients and suppliers will be to each other and to ensure the benefits of strategic alliances, a few strategies should be considered:

1. While requiring extraordinary performance in terms of value, speed and quality, most importantly, the agreement has to be good for both parties.
2. Facility managers should focus on the service component – exactly what the service is, what the needs are, and look at pricing issues last. It is important to know first what is needed from a vendor in order to call the relationship an alliance. What would the organization like to extract from the relationship in the future?
3. Installing employees on site with clients to be able to hold meetings to resolve issues that can't be successfully dealt with by phone or e-mail.
4. The greatest cost to consider is less related directly to the dollar and more to time and staff. These relationships require a company-wide commitment and they usually take years to build, requiring deliberate cultivation of senior management, and commitment from both sides.
5. Facility managers should consider their organization carefully. Does it have a culture conducive to forming partnerships, or does it prefer low-bid transactions? Do all departments agree?

References and bibliography

Alexander, K. (1996). *Facilities Management: Theory and Practice*. E. & F.N. Spon.

Atkin, B. and Brooks, A. (2000). *Total Facilities Management*. Blackwell Science.

Barrett, P. (1995). *Facilities Management: Towards Best Practice*. Blackwell Science.

Corbett, M. F. (1998). 'Outsourcing: beyond buying services'. *Facilities Design and Management*, 17(1), January, pp. 40-43.

Cotts, D. G. (1999). *The Facility Management Handbook* (2nd edition). AMACOM.
Damiani, A. S. (1998). *Moving up the Organization in Facilities Management: Proven Strategies to Increase Productivity in your Workforce*. Scitech Publishing.
Guzzo, R. A. and Dickson, M. W. (1996). 'Teams in Organizations: Recent research on performance and effectiveness'. *Annual Review of Psychology*, 47, pp. 307-38.
Hamer, J. M. (1988). *Facility Management Systems*. Van Nostrand Reinhold.
Hesselbein, F., Goldsmith, M. and Beckhard, R. (1997). *The Organization of the Future*. Jossey-Bass.
Hidaka, S. (1999). 'Facilities Management in Japan'. *Buildings*, 93(6,) June, p. 110.
Jurney, W. F. (1995). 'Outsourcing Non-core Functions Pays'. *National Underwriter* (Life/Health/Financial Services), 99(17), April, p. 18.
McGregor, W. and Then, D. (1999). *Facilities Management and the Business of Space*. Arnold.
Pacanowsky, M. (1995). 'Team Tools for Wicked Problems'. *Organizational Dynamics*, 23, pp. 36-51.
Park, A. (1994). *Facilities Management: An Explanation*. Macmillan.
Raiford, R. (1999). 'Into Uncharted Territory'. *Buildings*, 93(6), June, pp. 40-42.
Rondeau, E. P., Brown, R. K. and Lapides, P. D. (1995). *Facility Management*. John Wiley & Sons.
Salvetti, J. R. and Schell, N. D. (1995). 'Is Outsourcing Right for You?'. *Bank Marketing*, 27(10), October, pp. 45-47.
Sommerhoff, E. W. (1998). 'Working Together, Staying Together'. *Facilities Design and Management*, 17(10), October, pp. 60-62.
Stevens, D. (1997). *Strategic Thinking: Success Secrets of Big Business Projects*. McGraw-Hill.
Tompkins, J. A. (1996). *Facilities Planning* (2nd edition). John Wiley & Sons.

Continuity planning and disaster recovery

12.1 Introduction

According to one figure by the National Fire Protection Agency (US), 43 of every 100 businesses struck by a disaster never re-open, and 29% close after three years (Hawkins, 1998). Research by IBM showed that 80% of firms without relevant contingency plans which suffered a computer disaster went bankrupt within 18 months, with a further 10% going the same way within five years (Varcoe, 1994). Nonetheless, disasters can be managed and preparing a disaster recovery plan before one occurs is sound practice.

Most facility managers are responsible for emergency action plans during both working hours and after-hours. Problems can range from a leaking pipe to a full-blown disaster. Even routine events such as a power failure or equipment malfunction can approach natural disaster proportions for key operations such as computer systems if the situation is not handled properly. Often the solution requires multiple trades, all of which must be coordinated. If a facility emergency plan is not in place, then the alternative is to find the different vendors needed to cope with the disaster in the telephone directory. This may increase the chaos if the wrong people are hired (Pritscher, 1998).

'A thorough and well-designed emergency procedures plan enables the property manager to prepare before a disaster occurs, in order to minimize and perhaps prevent property damage and physical injury' states the Institute of Real Estate Management (IREM), Chicago, in DeMarco (1997).

Preparing an emergency procedures plan requires extensive research and teamwork and involves in-depth knowledge of the property and its residents, commercial tenants, the neighbourhood and the surrounding community (DeMarco, 1997). It is a time-consuming, labour-intensive task, however, the importance of such a plan cannot be overstated, and it is one of the most vital things that facility managers need to do. It is an example of risk management and includes all three aspects of risk identification, analysis and treatment.

12.2 Examples of disasters

Businesses must protect themselves from the damaging effects of manmade or natural disasters. This is not just an insurance issue, but one of continuity planning and disaster recovery. Examples of disasters include fire, earthquake, storm, flood, terrorist attack, explosion, power failure, equipment failure, vandalism and security breach. In recent times other 'disasters' have also been discussed, including environmental damage and Y2K problems.

12.3 Business loss reduction

A risk-reduction programme provides a cost-effective methodology for determining a company's risks for fire, earthquake, flood, terrorist attacks, power failures, etc., and systematically develops a plan of action to manage them. Emergency planning or a risk-reduction programme should outline short and long-term strategies for reducing business losses by incorporating contingency (continuity) plans and business recovery plans. These plans will provide a framework for company organization and operations after a disaster and should be kept current as facilities are modified, in accordance with the company's long-range plans and capital budgets. A successful risk management programme is essentially one of risk profiling, loss control and risk treatment. The key steps in a successful natural hazards risk management programme are:

1. Quantify the expected losses using engineering site assessments followed by portfolio analysis. This includes potential losses to buildings, all other assets, including key equipment and/or manufacturing systems (tools), business interruption and market share loss.
2. Maintain the programme, both through engineering site assessments of new acquisitions and through an annual portfolio re-analysis.
3. Conduct cost-benefit analyses to determine what loss control makes sense versus the purchase of insurance or other risk transfer mechanisms. In the cost benefit analyses, it is crucial to consider all major assets and the primary causes of business interruption and market share loss.
4. Strengthen, as determined by the cost-benefit analysis, those buildings, equipment, inventory and fire-protection systems that provide the company with the most return on its investment. To do this, the facility manager must take an active, or even leading role in the organization.
5. Enhance the risk management programme for future projects, including the development of design criteria beyond the normal building code requirements, third-party review of new construction, and diligent maintenance of loss control (Yanev and Conoscente, 1997).

12.4 Insurance

Any organization must have adequate insurance to cover all potential risks, and for the necessary values. Insurance may include public liability, employer liability (workers' compensation), building, contents and professional indemnity. The ones of particular interest to the facility manager are those that relate to the building and its contents.

While insurance may help to finance the reinstatement of facilities in the event of damage, the time this takes and the ability of the organization to remain 'in business' while repairs are enacted is also a consideration. Therefore as a supplement to insurance cover, organizations must look at minimizing disruption by effective continuity planning and disaster recovery. These are generally embodied in an emergency management plan.

12.5 General requirements of an emergency management plan

Human health and safety have always been the chief concerns during disaster recovery, but business continuity is also an important aspect of the process. To get operations back online quickly, facility managers have to understand how the business operates and determine what are the essential and non-essential functions. The best recovery plans recognize that business operations probably will resume in phases. The most critical operations that need to come back first are recognized by determining what would happen to each department if they couldn't access their information after certain periods of time (Tarricone, 1996).

To get started, or to review a current plan, three questions need to be asked:

1. Does the company have a facility emergency management plan in place which encompasses the information services and data centre, telecommunications, human resources, vital records, risk management, security, environmental concerns, product recovery and the facility itself?
2. Have procedures been established to protect essential records including computer files?
3. Is a single-source outside vendor (available 24 hours a day) part of the plan? A well-structured facility emergency plan will include a general contractor who has the contacts within the different trades, vendors and suppliers needed. One phone call to the general contractor triggers a chain reaction to deal with the situation (Pritscher, 1998).

Further to point 3, consider that disasters usually don't happen in a vacuum, and other companies may be affected by the same calamity. Facility managers should make sure their vendors have a disaster response plan and see whether or not they are going to be disabled by the same or another disaster (Tarricone, 1996).

The scope of the emergency management plan depends on the makeup of the facility or property and organization which in turn determines the extent of responsibility on the facility manager. However, regardless of who is leading the initiative, when preparing an emergency procedures plan the following guidelines may be considered (DeMarco, 1997; Pritscher, 1998):

1. Develop the emergency planning and management teams.
2. Identify potential emergencies.

3. Make sure safety is part of the facility emergency plan.
4. Prepare for an emergency; rehearse and test the plan. Test the facility team. Know the response time and where weaknesses exist and attend to them.
5. Adopt preventive measures.
6. Verify that all members of the team have the appropriate insurance to work at the facility and have supporting documents on file.
7. Make sure you or the general contractor has access to rental equipment and material suppliers, which can become scarce and hard to obtain in true emergency situations. In a true disaster, the demand for contractors increases dramatically and having immediate access from the outset may prove crucial.
8. Have it in writing, and then disseminate the plan to the tenants.
9. Respond to the emergency.
10. Debrief the emergency management team and others involved.

For the plan to work, the facility manager must be a decisive leader, with the ability to authorize an emergency response and direct an emergency management team.

12.6 The emergency management team

One of the first steps in preparing a disaster procedures plan is to assemble an emergency management team which should be comprised of two teams: primary and support. The primary team carries out the emergency plan and should consist of the facility manager, the onsite management staff, representatives from administration, and key maintenance staff. The support team should be comprised of specialists who may be called in for back-up guidance during an emergency, including contractors and suppliers, the architect, and utilities such as insurance, police and fire department (DeMarco, 1997).

Training the emergency team should begin with a tour of the property, pointing out relevant features such as the layout of the building, entrances and exits, roofs and stairwells, emergency equipment, communications equipment and life-safety equipment. Team members should know where to locate and operate utility switches, fire extinguishers and alarms, and emergency telephones. Each team member should be assigned specific duties, and should also know the jobs of the other team members in case of absenteeism. Everyone should understand all the procedures.

During an emergency, team members must remain in contact with the leader for instructions. Evacuation drills train team members to instinctively respond to emergencies and give the facility manager the opportunity to evaluate the emergency plan and make improvements before a real emergency arises. Building occupants need to be informed of the importance of practice evacuations so they do not regard them as a nuisance.

12.7 Creating an emergency procedures manual

The emergency procedures manual should contain reference information, directions for the emergency management team and management staff for each possible emergency, and directions for building occupants to safeguard people and property.

The reference information should include:

1. An up-to-date list of telephone numbers of team members, and emergency assistance resources.
2. General description of the building.
3. Local building codes and regulatory requirements.
4. Building systems information.
5. Floor plans and blueprints.
6. Insurance information.
7. A list of hazardous materials on the property. (DeMarco, 1997)

The manual should cover all emergencies that could happen on the property and the procedures that should be followed. For each emergency, the manual may address specific areas such as:

1. Detailed descriptions of the emergency management team's duties.
2. Flow chart of the chain of command.
3. Procedures to account for all employees and commercial tenants.
4. Evacuation and re-entrance procedures.
5. Reporting, documentation and regulatory procedures.
6. Restoration procedures. (DeMarco, 1997)

The facility manager, emergency management team members, and onsite managers each should have at least one copy of the emergency procedures manual. Additional copies may be stored in a safe place, in case originals are misplaced or destroyed.

12.8 The facility manager's responsibility

An organization's facility manager must always be in a position to respond to any disaster, meeting all the essential accommodation requirements within the response time constraints demanded by the operation. Precautions must be undertaken before the event to limit damage and to ensure that a fully co-ordinated proactive approach is taken. Disaster recovery planning must prepare for the worst scenario. Adequate preparation requires three key aspects to be addressed (Varcoe, 1994).

12.8.1 Risk identification

A detailed assessment of the risks to the integrity of the organization and its property needs to be undertaken and regularly reviewed as part of the provision of the normal security function. The risk resultant summary report should list an inventory of all buildings, equipment and systems; estimating their probable maximum losses for large and moderate disasters; and identify any particularly high-risk buildings or equipment that may cause lengthy business interruption in even relatively moderate events. The potential high-risk areas are prioritized so that management can decide whether to implement alternatives or alterations to reduce risk levels. They can establish a long-term strategy for risk reduction and control and develop implementation budgets. This information enables the corporate risk manager to determine insurance needs as well as to develop a long-term programme of loss control (Yanev and Conoscente, 1997).

12.8.2 Risk analysis

It is important to identify the demand by the establishment and collation of a reasonably detailed profile of the organization, together with its demand priorities for effecting a response. The information needs to be kept up to date so that the impact on the business function is immediately known.

The hierarchy of business function priority for each location needs to be established in co-operation with other senior members of the organization. The different components of the organization should be categorized relative to their impact on the essential function of the operation, for example:

1. *Priority One*: Essential core operation function critical to the performance of the organization's key activities (with immediate financial effects).
2. *Priority Two*: Important core operation function with medium- to long-term adverse financial effects.
3. *Priority Three*: Important business support function that represents a significant risk to the operation of a Priority One component by its absence.
4. *Priority Four*: Peripheral business support function that represents a low risk to a Priority One component, or a significant risk to a Priority Two, by its absence.
5. *Priority Five*: Other organizational components that represent little or no immediate risk to the fundamental operation of the organization by their absence.

The minimal operational requirements of each function must then be established, their essential core personnel must be identified and the minimum amount of space and equipment they need to perform their function must be determined. This establishes the current occupancy profile and the minimum accommodation demand profile of the organization. In the event of a disaster this provides the means of identifying who have been affected and in what priority order they need to be accommodated.

12.8.3 Risk treatment

The facility manager needs to take an active role in the preparation of a disaster recovery action plan. In this way plans can be made for the most efficient and speediest response in re-establishing the business.

The provision of the necessary accommodation requirements is an important measure. The first step is to assess the available space resources and match them to the demand profile. Information about all potential sources of accommodation should be kept up to date and accessible. This should address at least four potential sources as follows:

1. *Recovery of damaged areas*. Assure a quick response from suppliers by covering the event of a disaster within existing contracts (for example, cleaning, maintenance and security) for action if and when necessary. It also then provides the opportunity to clarify important issues such as contract procedures, guaranteed resourcing and response times, and cost, in advance.
2. *Available space within the existing estate*. If there are unaffected premises at the same or another location of an organization, all available space needs to be identified and assessed as a useful resource during emergency use (for example, the temporary decommissioning of common meeting, training or corporate hospitality facilities). A log of all space availability throughout the entire organizational estate needs to be kept up to date.
3. *Short-term temporary accommodation with existing resources*. The potential of other resources should also be considered, such as surface car parking areas to house mobile units. Another possible arrangement could be with mobile accommodation suppliers, which has the advantage of removing activity away from critical areas and thereby minimizing any delays in response.
4. *External resources within the local market*. Another possible source of space is the local office market. All available suitable properties should be constantly monitored, with perhaps one or two being 'selected' as preferred options at any particular time. Organizations might consider securing options on space at agreed rents while it remains available on the open market. The potential for home-working, based on a suitable IT and communications infrastructure, is another avenue of temporary accommodation provision. (Varcoe 1994)

With the space availability determined, the other major requirement to be met by the facility manager is the provision of the basic equipment necessities. The demand profile will again be identified in the impact assessment and the same principles of supply as outlined for space should be followed.

12.9 A practical implementation system

An effective response to a disaster requires the immediate availability of a lot of information co-ordinated into a cohesive document and circulated to all necessary management personnel. All critical staff should therefore have two copies of the document, for example, so that one can be kept at a separate, private location (Varcoe, 1994).

The first task following notification of the disaster will be to contact all the managers critical to the decision-making process and recovery period. After the emergency management team has appraised the nature and extent of the damage, they should consult the impact assessment report to determine the accommodation demand profile that needs to be provided. From this, an activity programme should be quickly produced, identifying when each affected organizational component can be brought back into operation. Then all members of staff through the organizational structure should be notified as soon as possible to allow key managers to put their action plans into effect and to tell those who are not initially needed to stay away until notified. The organization should also consider notifying customers and the public of the situation and indicate when everything will return to normal.

A comprehensive emergency management plan should be tested regularly. An effective emergency management plan certainly can lessen the potentially crippling effects of a major disaster. To ensure the plan is effective, consider the following:

1. Have all the divisions that could be part of the company emergency management team met together to identify virtually every possible disaster that could occur, what unique problems each would present and exactly how the company would respond to them?
2. Has a crisis communications chain of command been established that identifies who is responsible for communicating the event to the media and the public? Do other employees know what they can and cannot comment upon in such situations?
3. Have provisions been made to handle a rush of incoming phone calls from concerned relatives and others on any day and at any time?
4. Have back-up power sources, emergency lighting and communications equipment been strategically placed and tested frequently to ensure they do not fail when needed?
5. Has the disaster plan been tested more than just in theory, such as during mock disasters (Tillar, 1994)?

If the answer to each question is a ‘no’, chances are the company’s disaster preparedness is in need of improvement. These questions are only a few concerns that must be considered if a business is to be truly prepared to handle emergencies of any type. Even with the best plans, there inevitably will be numerous, unanticipated events during any major disaster. Without a solid plan, the likelihood of the business suffering serious damage is greatly increased.

Varcoe (1994) states that meeting the needs of an organization effectively in the wake of a disaster is all about ‘gaining time’ before the event. Every minute is invaluable and must not be wasted. When considering the provision of the accommodation need, it is essential to take a proactive approach. This approach facilitates the necessary provision at the earliest opportunity, and it also allows its achievement at a reasonable cost and not at all costs. Indeed, handled in the right way, an organization’s response to a disaster can paradoxically provide opportunities for gaining some beneficial effect. On the Monday morning after the Baltic Exchange bomb, Commercial Union placed full-page advertisements in most of the newspapers, showing a full-length photograph of their damaged headquarters captioned with the words, ‘business as usual; we won’t make a drama out of a crisis’.

12.10 Case study: Y2K preparedness

The ‘computer age’ has brought about the automation, networking, centralization and integration of facilities, plants, public and private infrastructures, communications, financial systems, health systems and just about everything else. Modern society depends upon computers for every facet of our social structure. The Year 2000 date problem caused by poor programming in software or anything containing a microprocessor or microcontroller was a source of potential global risk in the lead-up to 1 January 2000. Originally seen as just an information technology problem, affecting only the old legacy mainframe software programs, it was recognized to have far-reaching implications from the operation of desktop computing systems to the control and operation of facilities, plants, hospitals, finances, traffic lights, electrical power generation and transmission, water and sewage plants.

With over 10 billion microprocessors sold worldwide in the last five years alone (Hall, 1999), the estimated costs worldwide of fixing the problem ranged from the benchmark figure of the Gartner Group of $600 billion to TMR's $2 trillion (Anonymous, 1999). Buildings built since the 1970s are more reliant on embedded chips and as a result have the ability to malfunction.

Apart from the cost to fix the problem and the potential havoc reeked by such a glitch, all facility managers needed to be aware of the legal implications. If things go wrong, companies will want to point a finger – and they may well be pointing it at the facility manager (Anonymous, 1999).

While the issue is now history, and due to effective planning did not cause the possible havoc that may have been expected, it forms a useful example of the need for preparedness, quality control and reliability.

References and bibliography

Alexander, K. (1996). *Facilities Management: Theory and Practice*. E. & F.N. Spon.

Anonymous (1999). 'So you Thought 'TUPE' was Bad News: The millennium – a facilities management problem!'. *Facilities Management* (UK). June.
<http://www.cad.co.uk/fmuk/millennium.html>

Atkin, B. and Brooks, A. (2000). *Total Facilities Management*. Blackwell Science.

Barrett, P. (1995). *Facilities Management: Towards Best Practice*. Blackwell Science.

Cotts, D. G. (1999). *The Facility Management Handbook* (2nd edition). AMACOM.

DeMarco, A. (1997). 'Preparing for Disaster'. *Facilities Design and Management*, 16(1), January, pp. 42-43.

Hall, D. C. (1999). 'Implications of the Year 2000 Embedded Systems Problem'. *Facilities Management* (UK). June.
<http://www.cad.co.uk/fmuk/mill_impl.html>

Hamer, J. M. (1988). *Facility Management Systems*. Van Nostrand Reinhold.

Hawkins, B. L. (1998). 'When it Rains, it Pours'. *Facilities Design and Management*, 17(7), July, pp. 58-61.

McGregor, W. and Then, D. (1999). *Facilities Management and the Business of Space*. Arnold.

Park, A. (1994). *Facilities Management: An Explanation*. Macmillan.

Pritscher, C. (1998). 'Construction: Dial Once for Disaster Response'. *Facilities Design and Management*, 17(2), February, p. 26.

Rondeau, E. P., Brown, R. K. and Lapides, P. D. (1995). *Facility Management*. John Wiley & Sons.

Tarricone, P. (1996). 'Disasters Shouldn't Keep a Good Business Down'. *Facilities Design and Management*, 15(3), March, p. 19.

Tillar, M. J. (1994). 'Companies Need to Make Sure Disaster Plans are Current and Effective'. *Industrial Management*, 36(1), January/February, p. 1.

Tompkins, J. A. (1996). *Facilities Planning* (2nd edition). John Wiley & Sons.

Varcoe, B. J. (1994). 'Not Us, Surely?: Disaster recovery planning for premises'. *Facilities*, 12(9), August, pp. 11-14.

Yanev, P. and Conoscente, J. (1997). 'Coping with Quakes: Reactions'. *A Guide to European Risk Management Supplement*, 16-19 October.

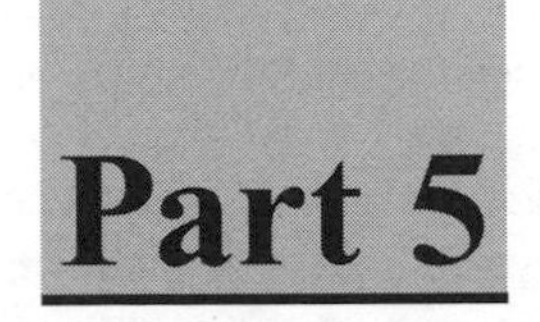

Property maintenance

All permanent facilities require maintenance. Maintenance is labour intensive, and in many parts of the world this means that maintenance work is expensive. It can occupy a reasonable proportion of the facility budget and yet in most cases there are insufficient funds to keep up the level of maintenance that otherwise might be thought appropriate. Facilities are regularly appraised and necessary maintenance is prioritized so that essential work is undertaken first.

But property maintenance is wider than just repairs. It includes cleaning activities necessary to maintain the facility in a habitable and healthy condition, energy for operation of building services and climate control, insurances, fees and other charges, repairs as before described, and of course continual replacement of building components. Many of these activities are regular (that is, daily, weekly, annual) and others are intermittent. While work is of a cyclical nature, it comprises a large number of different cycles that give rise to peaks in an expenditure cash flow.

The life cycle of a facility is commonly described as extending from design concept through to eventual demolition. This is known as the 'cradle to grave' life span during which a facility may have many owners and go through various metamorphoses before ultimate termination. A more meaningful time period is one that relates to the facility's use for a particular organization. In this case the period is known as the time horizon. Depending on owner or occupier motives, this may be quite a short period, or quite long.

Within this time horizon work must be carried out to keep the facility in what might be described as a normal or 'as new' condition. To achieve this, regular work must be undertaken. Buildings, like any asset, age and depreciate over time due to use, weathering and unplanned damage. Property maintenance concerns mitigating these effects so that both the value of the facility and its usefulness are maintained if not enhanced.

Maintenance planning is often seen as what facility management is about. But this is a fallacious view and is linked to purely operational or custodian concepts of building caretakers. Today facility management is a tactical and strategic professional discipline that has responsibility for ensuring that facilities are a productive support to the business activities that must be accommodated. Nevertheless, maintenance is an important matter and its proper planning requires considerable effort. Software solutions have been developed to assist in auditing the condition of a facility, prioritizing work and producing budget forecasts.

Because buildings age, it eventually becomes necessary to undertake substantial renewal work. This is not always related to a small residual value item, but could be a result of other criteria such as market expectations, fashion, legislation, new materials and functional change. When a facility is no longer suitable for the task it was intended to serve, it is said to be obsolete. This may trigger refurbishment or upgrade, adaptive use to a completely new purpose, sale or demolition. Many of these activities involve substantial construction and need to be effectively managed so as to minimize cost and limit unproductive business impacts.

Property maintenance is also held to include any support services that are entrusted to the facility manager. This may include matters such as car fleet management, parking, cafeteria services, childcare, health and fitness amenities, recycling systems, security, etc. Today many of these services are outsourced or shared with neighbouring organizations that are happy to work together in this way. Support services can have a positive effect on productivity and may be worth the cost and effort in some industries, particularly where it is difficult to gain and retain staff. High staff turnover will lead to increased training costs, organizational churn (disruption) and a loss of continuity that translates into lower productivity.

Collectively it is vital that facilities function smoothly and provide value for money to the organization. Facility cost represents a significant item on the balance sheet, yet is an area management would like to restrict so that capital can be directed to new high-return initiatives. It is an interesting tension that forms as a result.

This part deals with property maintenance. The matters discussed here have an operational focus, but include tactical and strategic issues as well. Chapter 13 looks at methods of maintenance planning and the systems that underpin the process. Chapter 14 covers obsolescence and refurbishment and the reasons behind premature renewal. Chapter 15 concludes with an examination of typical support services.

Maintenance is to facility management what measurement is to quantity surveying – a traditional and basic task, undeniably important, but increasingly being overwhelmed by contemporary value-adding activities that require higher problem-solving skills and professional judgement. Having said that, the success of the facility manager is all too often assessed by the little things that go wrong and the time it takes to fix them.

Maintenance planning

13.1 Introduction

Maintenance of physical assets, such as buildings, has long been a mainstream activity for facility managers. Although clearly an operational issue, the co-ordination, planning and budgeting of maintenance have been elevated to tactical, even strategic, levels to cope with the increased complexity of modern buildings and the pressure of keeping costs to a minimum. Maintenance may be undertaken within an organization by specialist staff or outsourced to contractors on an as-needs basis. Often a combination of the two is judged as the most appropriate solution.

Due to the need to regularize facility budgets, maintenance work usually has to be tailored to that which can be afforded. This means prioritizing work so that the most important or urgent tasks are completed first. Such an approach often leads to a reactive focus, which may result in planned work being deferred. However, putting off planned maintenance can accelerate the natural deterioration of building components and systems and lead to an increase in maintenance costs in the long run (Blanc, 1996).

Maintenance planning is a key activity for a successful organization. Facilities that work well may be taken for granted, but facilities with regular failure and complaint will be elevated into the corporate spotlight. The balance of available funds and essential work will always be present, and is a dilemma that the facility manager must regularly solve.

Maintenance is not restricted to work on the building fabric, contents, services and external works, but may also include other responsibilities such as supporting satellite work environments, information technology, software and specialist services like health facilities, food outlets, child care and car fleet management.

13.2 Types of maintenance

Maintenance varies in scope depending on the size of the facility and its age. Larger facilities obviously require more sophisticated management systems, and bigger annual budgets. Older facilities have higher unit costs due to the demands of careful restoration and the impact of any heritage regulations that may apply. Often facilities that incorporate historic or protected buildings require a higher proportion of in-house maintenance staff.

Maintenance can be reactive or proactive. Nevertheless, there are only three ways of maintaining buildings:

1. Corrective (unplanned) maintenance – the day-to-day work caused by unforeseen breakdowns, damage or emergency.
2. Predictive maintenance – planned repairs made on the basis of measured reductions in operating performance that herald future failure.
3. Preventative maintenance – planned repairs to restore elements or services to an acceptable standard, including routine cyclic work.

Cleaning is sometimes considered an integral part of maintenance, yet general cleaning of spaces and groundworks care are often outsourced completely and are rarely part of a typical maintenance plan. Other interpretations of cleaning are considered as mainstream maintenance, such as inspection and cleaning of air intake filters, high pressure water cleaning of façades, and clearing of debris from roofs and rainwater goods.

Tax implications are important in the definition of maintenance. Maintenance is normally deductible as an annual business expense, whereas capital improvement and renovation is not deductible from income, but rather depreciated over a number of years. Treating upgrades, alterations and churn as maintenance should be avoided from a cost reporting perspective.

Bernard Williams Associates (1999) suggest that in a modern air-conditioned office building, expenditure on maintenance of building services is 65% and expenditure on the building envelope is 25% of the total budget, leaving little for anything else. Building services includes mechanical and electrical services, communication, transportation (lifts, escalators, etc.), plumbing and security.

Many organizations employ at least one building services officer who has regular duties of looking after indoor air quality, temperature and essential equipment. Staff able to set up workstations and offices for new employees, rearrange space dividers and install new electrical and communication outlets are also common. Other major maintenance tasks are increasingly outsourced.

Pye (1999) found that, even in a well-developed industrial economy such as the UK, the majority of maintenance was still corrective in nature (over 61.3%). This was in comparison to preventative (34.0%) and predictive (4.7%). 'Many companies still treat maintenance as an overhead, external nuisance and necessary evil', says Richard Jones of MCP Management Consultants (cited in Pye, 1999).

13.3 Inspections

Planned maintenance may also be categorized as scheduled (undertaken to predetermined time intervals) or condition-based (undertaken following an inspection process). Due to the need to save money, most facility managers now use preventative condition-based assessment strategies (Bernard Williams Associates, 1999), yet the need to plan ahead for major work should not be forgotten.

An initial detailed condition survey and further regular inspections will enable an accurate anticipation of possible problems and allow a prioritized programme of maintenance work to be established (Warner, 1996). Emergency items receive the highest priority, including problems that will permit water entry (causing more damage) or issues of occupant safety, followed by matters which cannot be delayed without causing accelerated deterioration and increased expenditure. Usually, available funds will run out before all identified maintenance tasks are covered. Therefore work that can be deferred until a later time usually is.

Inspections are undertaken by experienced in-house staff. Their findings are documented and provide the basis of the maintenance plan. Each item of work is measured, costed and prioritized. Work is allocated commencement dates either within the current year or a following year. Software is often employed to record and sort proposed activities, and even to allocate in-house staff to tasks, manage external contractors, programme the work and record feedback for later analysis and ongoing planning.

13.4 Maintenance strategies

A properly considered maintenance policy should be based upon forward planning and preventative action (Warner, 1996). Planned maintenance is a question of doing the right thing at the right time, ensuring all parts of the building and its services are inspected and that appropriate maintenance is undertaken when necessary in an effective and cost-efficient manner.

Life-cost planning has emerged as one aspect of maintenance planning. Based on quantities of building work, forward estimates are made about when components are likely to need maintenance or replacement. Cleaning and energy costs are also part of this process. A long-term cash flow can be generated that helps to identify when large items of work are imminent. For example, replacement of built-up roofing is an expensive task and is not something that can be easily absorbed into an annual maintenance planning process.

A life-cost plan sets out targets for maintenance work, but this must be reviewed and interpreted in conjunction with actual performance and onsite inspections. A full inspection of the building should be undertaken every year to determine what needs to be done, and regular follow-up inspections may also be necessary. Identified work is costed and prioritized. Items of work that are not urgent can be deferred to enable budgets to be levelled across years.

The advantage of a life-cost approach is that it enables decisions to be made about items that are expensive to operate. It may be in the organization's interest to replace a piece of equipment prematurely if it can be shown that savings from implementation of new technology will result within a reasonable time period. Thus effective maintenance involves a continual reappraisal of facility performance in the context of new technology and changing business needs.

Maintenance policy must be linked to corporate goals and general business planning. The purpose of facilities is to support business, and therefore business decision processes must be transparent to the facility manager. Similarly, maintenance staff must be aware of future upgrade, refurbishment or relocation plans so that money is not wasted undertaking long-term repairs on assets with limited future application. For example, it would be of little benefit repainting office space walls when a major redesign of staff workspaces and purchase of new furniture was planned for the following year.

Proper training and updating of skills is a key issue for in-house maintenance staff. A good maintenance policy should make provision for skill upgrades and form the basis for decisions between direct employed labour and outsourced contractors. Customer satisfaction is an important part of the facility mission statement, and requires educational programmes that instill appropriate ideals in staff and provide the ability to handle conflict if it should arise.

13.5 Effective planning

Maintenance strategies are codified into formal policies that underpin maintenance operations. Such policies are intended to reflect the organization's approach to maintenance and delineate responsibility between business units. Effective planning can be judged by reference to agreed performance indicators such as:

1. Response to emergency call-out.
2. Down time of essential plant and equipment.
3. Physical inconvenience to building occupants.
4. Quality of work carried out.
5. Cost. (Bernard Williams Associates, 1999)

Most organizations are looking to reduce costs, and often maintenance of facilities is one of the areas that are targeted for reform. It is therefore crucial that the performance of facilities are quantifiably linked to worker productivity, and that allocated money is seen as delivering tangible improvements to overall business performance. Such a focus elevates maintenance from a necessary operational cost to one of strategic importance.

Maintenance costs need to be correctly allocated and recorded. Areas of high expense need to be identified and the reasons for such over-run examined. A database of maintenance work should include information about the reason for the work, who reported it, when it was started, finished and checked, and who undertook the work. Quality assurance can lead to increased satisfaction by building occupants and improve the perception that facilities are being properly managed. Minor problems can soon turn into major confrontations, lead to occupant dissatisfaction and reduce overall productivity.

13.6 Computerized maintenance management systems

Computerized maintenance management systems (CMMS) are now common in large owner-occupied facilities. They enable an integrated approach to planning, financial control, administration and feedback that can improve efficiency and increase satisfaction levels within an organization.

At a basic level, a CMMS collects information about what is going on and then applies simple rules to arrive at decisions (Pye, 1999). Properly used, it makes for better maintenance decisions in real-time, and helps keep everything up and running as inexpensively as possible (Gould, 1998). It can advise on what maintenance is essential for meeting production targets, and what is just routine. It can also store and manage the information needed to improve facility performance.

Serious consideration must be given to the selection and design of a CMMS. The level of process discipline, complex support systems and organizational capabilities required for the application of technology is often underestimated. The selection process for CMMS software must focus on its intended purpose and be directly related to achieving reliability at lowest cost. One of the biggest causes of failure in CMMS implementation is that little thought is given to the manner in which the software will be used or to the effort required to operate it successfully (Cooper, 1998).

Given appropriate management expertise, the facility management functions that can be performed by a CMMS include:

1. Planned (preventive) maintenance, which includes indoor air quality via planned maintenance of air handling units, emergency generator testing, fire alarm testing, fire extinguisher testing, smoke control testing and monitoring/measurement of these items to ensure efficient facility management.
2. HVAC automation systems, which includes smoke control systems exercised, indoor air quality monitoring/measurement, self-diagnostic testing for accuracy of devices, criteria/performance testing for speciality rooms, run-times for initiating predictive maintenance work orders and monitoring/measurement of these items to assure efficient facility management.
3. Fire alarm systems, which includes self-diagnostic testing, annual testing and monitoring/measurement of these issues to ensure efficient facility management (McKew, 1998).

The integration of CMMS with the overall building management system can take the above data and convert it into regulatory compliance for life safety and environmental issues. In addition, there are other management requirements, such as customer satisfaction, that should be monitored and measured for quality assurance.

A CMMS can be promoted to the level of an enterprise by integrating it with an enterprise resource planning (ERP) system. There are many benefits that can flow to an organization if a CMMS is integrated into an organization's overall management information system. An ERP system is a fully integrated software package that usually tackles the mainstream parts of a business such as finance, production and logistics. When a CMMS is part of an ERP, a host of useful functions is available.

One of these new functions is improved preventative maintenance. The benefits of integrating a CMMS occur in eight main categories (Cooper, 1998):

1. *Logistics.* Integration with the corporate inventory and procurement systems would allow carrying out of maintenance stock control and purchasing by cor-

porate systems and would avoid the overheads of using separate software for the maintenance unit. The integration with the CMMS must ensure that all stores and purchasing costs are allocated to work orders in the CMMS, and also allocating them to the life-cycle costs of the equipment.

2. *Financial management.* Collecting data to transfer maintenance costs to the financial systems would allow allocating maintenance costs to production cost centres and help to establish the true cost of production and profit contribution for each part of the process. In addition, the equipment records should be linked with the fixed asset database to establish an accurate register of the company's assets.
3. *Human resources.* Some critical maintenance tasks are highly complex and require specially skilled people. Checking that an appropriately trained person is assigned to these tasks is simplified if the CMMS is integrated to human resources software. Other areas of integration with human resources systems include checking vacation records when planning maintenance work.
4. *Executive information systems.* To complement the recording and storing capabilities of detailed transactions of maintenance work, a CMMS needs to be integrated with specialized executive information or decision support systems, to create tailor-made reports and inquiries. Many organizations will permit reports, charts, etc., of established key performance indicators of maintenance management to be published on their local network so that it is accessible by suitably authorized people using a web browser. ERP systems can present the data and include information gathered from other systems, such as production information.
5. *Production planning.* Most textbooks on maintenance planning urge readers to plan maintenance and production work together, to utilize production 'windows' and to maximize the availability of equipment to the production process. Some ERP suppliers have maintenance work orders and production order records using the same format, therefore allowing the same program to be used for displaying and planning both types of work simultaneously on the same screen.
6. *Condition monitoring.* Systems that analyse data from condition monitoring sensors, such as vibration analysers, usually signal an alarm condition when one is detected. Ideally, the CMMS should export to the condition monitoring system details of the operations required to take the condition monitoring readings, then receive back data on any alarm conditions detected and create maintenance requests automatically to investigate the alarm conditions and take appropriate corrective action.
7. *DCS and SCADA*. By designing distributed control systems (DCS) and supervisory control and data acquisition (SCADA) systems to include sensors required for maintenance, such as meters, temperature alarms and vibration analysers, total predictive maintenance environments can be set up where maintenance, production and design departments work together.
8. *Time and attendance.* ERP permits recording the number of hours worked by an employee via a keyboard through the time clocks used by an organization's time and attendance system. ERP systems usually have a standard time and attendance module, so integration with the CMMS module would be easy to provide.

A CMMS is particularly important for industrial and engineering facilities with large equipment content where equipment failure directly leads to downtime and lost production.

13.7 Conclusion

Maintenance is an indispensable part of any facility management operation. Its effective planning and execution can deliver tangible benefits to an organization. Large organizations require sophisticated software solutions to assist in maintenance management, yet this often must be achieved as budgets come under increasing pressure from corporate cost-cutting and down-sizing in an attempt to become more efficient and competitive.

The facility manager must balance necessary repair with available funds, and ensure reliable infrastructure in the context of continual change. Maintenance has been elevated from a purely operational level to one with both tactical and strategic implications. Taking a long-term view of facilities and keeping abreast of business needs are two important characteristics that will help to ensure maintenance is both effective and timely.

References and bibliography

Alexander, K. (1996). *Facilities Management: Theory and Practice*. E. & F.N. Spon.

Atkin, B. and Brooks, A. (2000). *Total Facilities Management*. Blackwell Science.

Barrett, P. (1995). *Facilities Management: Towards Best Practice*. Blackwell Science.

Bernard Williams Associates (1999). *Facilities Economics*. Building Economics Bureau Limited.

Blanc, A. (1996). 'Maintenance of the Building Structure and Fabric'. In *Building Maintenance and Preservation: A Guide to Design and Management* (E. Mills, ed.), Architectural Press, pp. 88-108.

Cooper, C. (1998). 'An Integrated System for Mill Maintenance'. *Pulp and Paper International*, 40(12), December, pp. 19-21.

Cotts, D. G. (1999). *The Facility Management Handbook* (2nd edition). AMACOM.

Gould, L. S. (1998). 'Keeping up, Running and Profitable with CMMSs'. *Automotive Manufacturing and Production*, 110(8), August, pp. 68-71.

Hamer, J. M. (1988). *Facility Management Systems*. Van Nostrand Reinhold.

McGregor, W. and Then, D. (1999). *Facilities Management and the Business of Space*. Arnold.

McKew, H. J. (1998). 'Unite CMMS and BAS to Create Compliance'. *Facilities Design and Management*, 17(5), May, pp. 66-68.

Park, A. (1994). *Facilities Management: An Explanation*. Macmillan.

Price, S. (1997). 'Facilities Planning: A Perspective for the Information Age'. *IIE Solutions*, August, pp. 20-22.

Pye, A. (1999). 'Can CMMS Resolve Work Scheduling Conflicts?'. *Works Management*, 52(5), May, pp. 54-57.

Rondeau, E. P., Brown, R. K. and Lapides, P. D. (1995). *Facility Management*. John Wiley & Sons.

Tompkins, J. A. (1996). *Facilities Planning* (2nd edition). John Wiley & Sons.

Warner, D. L. (1996). 'Maintenance Policy, Programming and Information Feedback'. In *Building Maintenance and Preservation: A Guide to Design and Management* (E. Mills, ed.), Architectural Press, pp. 195-201.

Obsolescence and refurbishment

14.1 Introduction

Buildings are major assets and often form a significant part of facility management operations. Although buildings are long lasting they require continual maintenance and restoration over their lives. Eventually, most buildings become inappropriate for their original purpose due to obsolescence, or can become redundant due to change in demand for their service. It is at these times that major change is likely, such as demolition to make way for new construction, or some form of refurbishment.

Refurbishment can of itself take many forms, ranging from simple redecoration to major reconstruction. Sometimes the buildings are in good condition but the services and technology within them are outdated, in which case a retrofit process may be undertaken. In other cases a particular function may be no longer relevant or desired, and buildings may be converted to a new purpose altogether. This is called adaptive reuse.

Older buildings often have a character that can significantly contribute to the culture of a society and conserve aspects of its history. The preservation of these buildings is important and can have many advantages to their owners. Facility managers are frequently faced with decisions about whether to rent or buy, whether to extend or sell, and whether to refurbish or construct. Usually these are financial decisions, but there are other issues that can bear on the final choice, like suitable corporate image, continuity of business operations and relocation (including promotion) expenses.

Relocation emerges as a major event for large organizations, and one that requires extensive planning. What is at issue is ensuring that business operates in a smooth and uninterrupted fashion before, during and after the move. Relocation is involved whenever premises are changed or altered, including dealing with the effects of corporate expansion and downsizing.

For some organizations, the need for physical facilities is decreasing. Technological innovations, like e-commerce, can have significant implications for the way businesses operate and their associated demands for space. For example, a retail organization involved in selling toys may convert from physical stores and over-the-counter sales, to Internet purchasing and delivery direct from a warehouse. In this case, premises close to other retailers are no longer relevant, and a large central warehouse near major highways may be more appropriate.

What is certain is that change is always at hand, and facility managers are change agents for their organizations and must ensure that the infrastructure they manage is constantly in tune with overall business goals.

14.2 Obsolescence

Johnson (1996, p. 209) indicates that, as society has advanced, its use of buildings has become more temporal. He states that 'advances in technology and commerce, including the growth of industrial and office automation, and user demands for more comfortable environments for work and leisure have led to large numbers of buildings becoming obsolete or redundant and these changes have provided an abundance of buildings suitable for rehabilitation and reuse'.

The effective life of a building or other asset in the past has been particularly difficult to forecast because of premature obsolescence (Seeley, 1983). This may be caused by one or more of the following:

1. *Physical obsolescence.* The physical life of the building is the period from construction to the time when physical collapse is possible. In reality, most buildings never reach this point as they are demolished or refurbished for other reasons.
2. *Economic obsolescence.* The economic life of the building is the period from construction to economic obsolescence, that is, the period of time over which occupation of a particular building is considered to be the least cost alternative for meeting a particular objective.
3. *Functional obsolescence.* The functional life of the building is the period from construction to the time when the building ceases to function for the same purpose as that for which it was built. Many clients of the building industry, particularly in manufacturing industries, require a building for a process that often has a short life span. Functional obsolescence can also include the need for locational change.
4. *Technological obsolescence.* This occurs when the building or component is no longer technologically superior to alternatives and replacement is undertaken because of expected lower operating costs or greater efficiency.
5. *Social obsolescence.* Fashion changes in society can lead to the need for building renovation or replacement.
6. *Legal obsolescence.* Revised safety regulations, building ordinances or environmental controls may lead to legal obsolescence.

Therefore buildings can become obsolete long before their physical life has come to an end. Some highly specialized buildings can be obsolete within ten years. It is generally regarded that the effective life of modern office buildings is 20–25 years. It is wise to design buildings for change by making them flexible and modular yet with sufficient structural integrity to support alternative functional use.

14.3 Redundancy

Redundancy is not the same as obsolescence. Buildings can become redundant when they are superfluous to their owners, even though they may be attractive to other potential owners. Redundant assets can either be modified to a new purpose, sold or leased.

Redundancy can occur within buildings as well as for total buildings. For example, a particular treatment facility in a hospital may no longer be needed, and this space is therefore available to be reused for some other purpose. Redundant space is a wasteful use of resources.

14.4 Refurbishment

Refurbishment includes minor work like redecoration of wall, floor and ceiling surfaces, new fit-out, or space rearrangement, or more significant activity like façade upgrades, retrofitting of services, major reconstruction, extension and conversion to new functional use. Major refurbishment is undertaken by building owners to make facilities more productive or to attract higher rental income. Minor refurbishment can also be undertaken by tenants to fit-out their spaces to attract customers and to create a new corporate image.

In addition to the growing availability of obsolete or redundant buildings, a further benefit in favour of their rehabilitation is that many older buildings were soundly constructed using high quality materials, forming a suitable basis for restoration and improvement. But there are many other advantages of rehabilitating older buildings over demolition and construction of new space. These can be generally categorized as economic, environmental and social benefits.

14.4.1 Economic benefits

A major benefit is that rehabilitated space can be created more quickly than new space, unless perhaps extensive structural reconstruction is required. Johnson (1996) suggests that rehabilitation typically takes half to three-quarters of the time necessary to demolish and reconstruct the same floor area. The shorter development period reduces the cost of financing and the effect of inflation on building costs, and the organization can begin earning revenue faster, reducing temporary accommodation expenses.

Despite the time advantages, the cost of converting a building is generally less than new construction because many of the building elements already exist. Given there are no expensive problems to overcome, like asbestos removal or foundation subsidence, the reuse of structural elements is a significant saving. However, older buildings often do not comply with present regulations, particularly in the area of fire safety, which may generate some structural changes or additional protective measures.

It is essential that any building being considered for major refurbishment have a thorough survey undertaken to confirm its structural and constructional quality, and its compliance with building ordinances.

14.4.2 Environmental benefits

Environmental benefits from rehabilitation arise through the recycling of materials, reuse of structural elements and the reduction in generated waste. These translate into cost advantages to the owner, but have much wider environmental implications as well. Older buildings typically were constructed using quality materials with lives well in excess of modern counterparts.

Furthermore, many older buildings employ massive construction in their external envelope, which can reduce energy consumption in heating and cooling and deliver long-term operational efficiencies. Opening windows, natural ventilation and natural lighting are all desirable qualities where external noise and pollution is not an issue. Low-rise structures can also eliminate the need for expensive vertical transportation systems, such as lifts.

The reuse of existing public infrastructure, like telecommunications, water, gas, sewerage and drainage, can relieve demands on local authorities to extend infrastructure and to reclaim natural landscapes for sprawling urban development.

14.4.3 Social benefits

Older buildings also provide social benefits. They retain attractive streetscapes, add character, and can provide status and image to an organization through the use of massive and highly crafted materials. Older buildings are often in advantageous locations in city centres and close to transport. They also add to a sense of community and are frequently appreciated as comfortable working environments by their occupants.

However issues of legislative compliance, fire safety, disabled access and heritage constraints (such as a requirement for façade retention) are possible disadvantages that should be properly explored.

14.5 Adaptive reuse

Adaptive reuse is a special form of refurbishment that poses quite difficult challenges for designers. Changing the class (functional classification) of a building will introduce new regulatory conditions and perhaps require zoning consent. There are clear economic, environmental and social benefits that can make this option attractive to developers. In some cases increases in floor space ratios can be obtained and concessions received for pursuing government policy directions by regenerating derelict public assets. In recent years redundant city office buildings have been converted into high quality residential apartments, bringing people back to cities and in the process revitalizing them.

Adaptive reuse can be quite dramatic. For example, conversion of disused industrial factories into shopping centres or churches into restaurants is possible. Facility managers should be conscious of adaptive reuse solutions to redundant space and continually think about more productive uses for existing premises.

14.6 Relocation

The decision to relocate can result from a wide range of pressures such as company growth and change, shortage of space, financial constraint, organizational inefficiency, the need for rationalization (centralization) or decentralization, and core operational problems. Generally the motivation can be categorized as corporate led (resulting from business factors) or premises led (resulting from accommodation factors), and depending on the motivation the facility manager will have a different range of issues with which to deal (Bernard Williams Associates, 1999).

Relocation may be temporary while refurbishment activities are underway, or permanent to a new building constructed or rented elsewhere. In either case it is wise to engage professional expertise to plan and execute the move. Facility managers who unintentionally shortcut the move process in an attempt to keep costs down often discover it costs more than they budgeted by the time the move is complete (Kemble, 1999). Distracting staff from more productive pursuits will create unseen costs. Proper planning is vital if business processes are to be uninterrupted.

The decision to relocate is one of strategy; the planning of how to do it is one of tactics; and the move itself is one of operation. This activity therefore illustrates the three-tier attributes of facility management and suggests that the input of the facility manager is more valuable for the strategic decisions, leaving operational solutions as a potentially outsourced task.

Relocation can range from a total corporate move to another piece of internal churn. For major relocations, the following issues need to be addressed and resolved:

1. The budget.
2. Staff or facilities requiring special treatment.
3. Lead times for installation of services like security, telephones, IT networks, etc.
4. Permissions and licences.
5. Equipment or substances requiring special handling.
6. Design, production and delivery of new stationery.
7. Advice to customers, suppliers and other stakeholders of the move date, new address, etc. (Bernard Williams Associates, 1999).

Weekend or out-of-hours moves are less disruptive. Employees may be asked to pack their personal belongings into labelled boxes on a Friday afternoon, and on Monday morning they can unpack their belongings in the new location. All equipment and supplies will be transported over the weekend, and in effect it can be 'business as usual'. However, some organizations may require 24-hour operation, and a staged move may therefore be appropriate.

Relocation offers an opportunity for process re-engineering and a reassessment of space needs and relationships. Rather than transporting an existing operation wholly to a new premises, designing the new accommodation to deliver productivity improvements is worth the time and effort.

14.7 Conclusion

With the increasing pace of change, modern purpose-built facilities will become obsolete more quickly. Increased environmental concerns may result in decisions to closely tailor new facilities to their first owner as an unnecessary extravagance. The potential for building adaptation to different uses in the future may become an important issue for designer and clients alike. Pressure for greater conservation and careful use of resources makes it logical to view existing buildings as a valuable asset, worthy of preservation and concern. In this way society may return to a philosophy of more sparing use of materials, optimum deployment of available skills and in-built adaptability – a long-life, loose-fit, low-energy approach of the type that characterized building design in the pre-industrial age (Johnson, 1996).

References and bibliography

Alexander, K. (1996). *Facilities Management: Theory and Practice*. E. & F.N. Spon.

Atkin, B. and Brooks, A. (2000). *Total Facilities Management*. Blackwell Science.

Barrett, P. (1995). *Facilities Management: Towards Best Practice*. Blackwell Science.

Bernard Williams Associates (1999). *Facilities Economics*. Building Economics Bureau Limited.

Cotts, D. G. (1999). *The Facility Management Handbook* (2nd edition). AMACOM.

Johnson, A. (1996). 'Rehabilitation and Reuse of Existing Buildings'. In *Building Maintenance and Preservation: A Guide to Design and Management* (E. Mills, ed.), Architectural Press, pp. 209-230.

Kemble, L. (1999). 'Budget Improvement into your Move Plan: The key to a successful relocation effort is to properly plan at the conception of the project'. *AFE Facilities Engineering Journal*, December. <http://www.facilitiesnet.com/fn/NS/NS3a0aa.html>

McGregor, W. and Then, D. (1999). *Facilities Management and the Business of Space*. Arnold.

Park, A. (1994). *Facilities Management: An Explanation*. Macmillan.

Rondeau, E. P., Brown, R. K. and Lapides, P. D. (1995). *Facility Management*. John Wiley & Sons.

Seeley, I. H. (1983). *Building Economics: Appraisal and Control of Building Design Cost and Efficiency* (3rd edition). Macmillan.

Seeley, I. H. (1987). *Building Maintenance* (2nd edition). Macmillan.

Stevens, D. (1997). *Strategic Thinking: Success Secrets of Big Business Projects*. McGraw-Hill.

Tompkins, J. A. (1996). *Facilities Planning* (2nd edition). John Wiley & Sons.

Support services

15.1 Introduction

Modern premises contain a range of support services that should fall within the area of responsibility of the facility manager. In some larger organizations these services may be discrete operations under separate management divisions, or even outsourced to external providers. It is not uncommon for large organizations to provide services like childcare, health and gymnasium programmes, restaurant, library, transportation and social/sporting clubs, all of which require specialist facilities.

This chapter identifies some of the key support services, not to describe their operational requirements, but rather to highlight the impact they have for improving productivity and increasing employee (or customer) satisfaction. There is a strong case for all of these services to come under the direction of the facility manager, although this is not always the reality.

15.2 Non-core activities

Non-core support services comprise both physical facilities and delivery processes. Their nature varies from organization to organization and may comprise a mix of in-house and outsourced strategies. Typical support services are discussed in the context of their contribution to business productivity.

15.2.1 Security

Security is one of the more universal support services that must be provided. In some cases this service is so important that it warrants its own management structure and financing. Nevertheless there is a case for it being part of the overall facility management process, particularly where the service is outsourced. Installation and maintenance of security equipment obviously requires considerable co-ordination.

Security can take on many aspects. It can relate to protection of premises, equipment and other assets, personnel, data, communication systems, cash transfers and public relations. In each case the threat can be both internal and external to the organization (Bernard Williams Associates, 1999).

The establishment and management of access control to premises and within premises and the separation of staff areas from public areas is increasingly necessary. Solutions to these issues may require help desk or reception facilities, special equipment and database management processes, in addition to security personnel. Continuity and disaster recovery and the response to emergency situations are also important roles for security.

Training of security staff and the establishment of policies are instrumental to efficient operational performance. However, other staff also need to be aware of procedures for access, parking, visitors, etc., and what to do in an emergency. Online information can assist in disseminating instructions to building occupants, but activities like regular fire drills are still an important part of human safety.

An effective security system is vital for organizations to perform at their best and to provide a safe working environment for staff.

15.2.2 Information technology

Information technology and facility management often co-exist, yet there is a strong argument for combining these two management divisions under one leadership model. Clearly there is overlap in the provision of infrastructure, but the reason these areas may be separated is usually a result of diversity in personnel background, knowledge and education. Traditionally, facility managers have not had IT training or experience, and similarly IT people have had little understanding of property and facility operation. However, if facility managers can improve their understanding of IT strategy and support, significant advantages will flow from the synergy that is created. Both share a common goal of 'customer satisfaction'.

Worker productivity is a key issue. Improving business efficiency is usually a function of technology provision combined with flexibility of space usage. It is difficult to achieve an integrated solution when two separate management divisions with separate budget allocations and priorities are involved. A streamlining of the organizational hierarchy can enable more innovative methods of supporting business to be conceptualized, proposed and implemented.

Information technology comprises the provision of equipment (hardware), appropriate tools (software) and the infrastructure to enable people to work collaboratively (network), as well as the personnel to provide hands-on support. Other issues such as financing, data storage and back-up, website development, Internet and e-mail access, equipment upgrade policy, software selection and licence agreements, security and virus protection all involve important decisions.

Intelligent buildings are a further example of the potential synergy between traditional facility management and IT support processes. Building management systems, computer-aided facility management systems (including CAD), and integrated enterprise resource planning systems are now enabling the facility manager to be more aware and involved in IT developments, which in the future will increase the likelihood that all business infrastructure support will be centralized under one line of management.

15.2.3 Communications

Communications involves information technology and database sharing, but also encompass other forms of interaction such as mail delivery, document retrieval, telephone, facsimile, conferencing (audio and video) and client presentation. The facility manager is usually responsible for the provision of infrastructure to support such communications.

The greater use of electronic forms of communication and the ways in which this can transform traditional work practices and improve productivity is now obvious. Team-based processes can effectively be enhanced using groupware software solutions that establish project and corporate databases, co-ordinate diaries and calendars, timetable space needs and provide staff directories. Again there is a merging of traditional facility management, human resource management and information technology functions that can assist intra- and inter-office communications, as well as links with the global marketplace. Satellite offices, telecommuting, hot-desking, hotelling and other modern workplace solutions are totally reliant on an effective communications strategy. Mobile phones, pagers and e-mail accounts are now indispensable corporate tools.

Organizations must have a clear communications strategy and provide the right tools to people to support their work tasks and patterns. The facility manager is a key player in the development of this strategy, as it will have far-reaching effects on the ways in which facilities are conceptualized and delivered.

15.2.4 Reprographics

The much-heralded day of the 'paperless office' has never really materialized. In contrast, organizations typically consume more paper than ever before. Convenience, legal responsibilities and high presentation standards have ensured that paper communication is still mainstream. The only real advance in this area is the use of electronic mail (e-mail) to substitute for fax and post, but even this represents only a small proportion of total communications.

Some organizations require their own in-house printing facilities that can produce marketing information, product packaging, advertising and other important support functions. These facilities have expensive equipment that must be maintained and upgraded just like other business assets. All organizations use photocopiers and desktop printers to reproduce documents.

The purchase and processing of paper products is still a major cost for business. Strategies are necessary to ensure recycled products are used wherever possible, and to ensure that waste paper is similarly recycled for future purposes. An effective strategy needs to be developed to reuse as much paper as possible, and to collect discarded paper for recycling. Not only does such a strategy have environmental benefits, but there are also large cost savings that can be realized.

15.2.5 Filing and storage

Space for filing and storage is usually at a premium. Systems are available to compact the amount of floor space necessary to hold records. While some storage is electronic, the majority of business records are still printed as hard-copy for tax and auditing purposes. Some documents must be kept for many years.

Microfilming is still a valid way of storing large volumes of archival documents. But these days digital images of documents are more common, and have the advantage of being electronically indexed and printed on demand. CD and DVD technologies are used to store the digital images efficiently.

Worker productivity suggests fast access is required to information, preferably online and from any location. The solution is a combination of conventional paper handling solutions and innovative information technology. The facility manager should be aware of storage implications for facilities and provide appropriate support and security.

15.2.6 Furniture and equipment

An important business support function is the timely provision and maintenance of furniture, equipment and other movable assets. A database is normally used to record serial numbers, locations, costs, maintenance requirements, dates of purchase, warranty information and the like. Open-plan or closed office layouts for administrative functions will dictate different furniture and equipment solutions. Flexibility and modularity remain vital attributes for selection.

Furniture must have proper ergonomic design with adjustable components to suit people of different sizes and physical characteristics. Choice of furniture should also be based on the task for which it will be used, rather than the rank of the person who will use it. Selection choices that can minimize churn costs as employees move within organizations is an important consideration.

While the open-plan office has numerous advantages in terms of flexibility and space efficiency, it suffers from a lack of visual and noise privacy. In such cases the use of dividing screens and appropriate finishes on floors, walls and ceilings is critical. But the furniture and equipment that reside in open-plan areas must also be chosen to minimize noise and vapour emission, and be sited in places that cause the least disturbance to adjacent tasks.

The introduction of computers into the workplace has dictated certain standards of workstation size to accommodate equipment, in particular the computer screen. New flat-screen technology is changing those standards, enabling workstations to be smaller. Yet simultaneously, voice-activated technology and mobile phones are increasing office noise levels and decreasing the effectiveness of existing open-plan arrangements.

15.2.7 Cleaning

Routine cleaning is commonly an outsourced activity. This, however, does not mean that the quality of the task is unimportant. Hygienic and dust-free environments are critical to the health of building occupants, to reduce sick leave and absenteeism and therefore increase productivity. Cleaning is normally undertaken out-of-hours so that disturbance to core activities is minimized.

Bernard Williams Associates (1999) identify the following principal cost drivers relevant to cleaning:

1. The labour rate.
2. The nature and frequency of cleans.
3. Adequate storage and access to equipment and supplies.
4. Scope and quality specification.
5. Health and safety provisions.

Some facilities have spaces that require high levels of cleanliness, perhaps requiring the use of special ventilation and extraction systems. Facility managers should also be conscious of the need to provide adequate safety for dangerous tasks, like external façade cleaning, and to enable access to secure areas. Waste handling and disposal is also critical to minimize health and fire hazards.

15.2.8 Stationery and business supplies

The provision of stationery is also a task that may fall within the responsibility of the facility manager. While it may seem a low status activity, the cost of stationery can be surprisingly high. Most of the cost is related to paper supply. Adequate stocks of key resources and controlled access must form part of its effective management.

Stationery also includes equipment supplies, like toner and ink cartridges, personal effects like filing systems and storage bins, writing implements, computer media and presentational tools.

15.3 Other support services

Other services may include car fleet control, catering, landscaping, childcare, accommodation, porterage, reception, medical and healthcare, records management, library resources, etc. Whether these fall within the purview of the facility manager largely depends on how the organization has evolved in its allocation of managerial responsibility.

Specialist facilities that represent non-core business activity may be considered an unnecessary expense. While they cost money to operate, or at best break even, they provide benefit through gaining and retaining staff. Childcare facilities, for example, enable primary care providers to work part-time or full-time without the distractions of supervising or entertaining children. Health facilities can relieve stress and improve the performance of employees without lost time travelling to remote facilities.

The economics of specialist facilities requires careful consideration and analysis. In hard times they might be services that are cut back, but their impact on the morale and general satisfaction of staff may be very important. It can be hard to quantify these benefits and to weigh them against tangible monetary criteria.

Some specialist facilities, like food preparation and transportation management, serve important corporate functions. For example, a hospital without in-house catering facilities would be unthinkable. Depending on the organization, facility managers will need to manage a range of support services that are peripheral to the main business function.

15.4 Customer satisfaction

Regardless of the type of support service provided, and whether it is supplied using in-house or external resources, the quality of the service needs to be evaluated and subject to continual process improvement (Kennedy, 1998). Customers may not necessarily be restricted to employees, but include visitors, clients, consumers, suppliers and other stakeholders. Regular review of performance is the responsibility of the facility manager, and suitable mechanisms need to be put in place to collect and analyse user opinion.

15.5 Conclusion

There are a diverse range of support services that organizations may require to function efficiently. Proper planning, resources, accounting, maintenance and review of these services is vital. The facility manager is an appropriate person to oversee all aspects of business infrastructure and non-core functions. However, it is equally vital that management of support services is not interpreted as a purely operational task, but one with both tactical and strategic implications.

An interesting advance in strategic facility planning is the sharing of support services between organizations. This may involve facilities like swimming pools, restaurants or childcare, or processes like security, hospitality and information technology. Large premises with multiple leased spaces lend themselves to this approach and can produce economies of scale over individualized management.

References and bibliography

Alexander, K. (1996). *Facilities Management: Theory and Practice*. E. & F.N. Spon.

Atkin, B. and Brooks, A. (2000). *Total Facilities Management*. Blackwell Science.

Barrett, P. (1995). *Facilities Management: Towards Best Practice*. Blackwell Science.

Bernard Williams Associates (1999). *Facilities Economics*. Building Economics Bureau Limited.

Cotts, D. G. (1999). *The Facility Management Handbook* (2nd edition). AMACOM.

Damiani, A. S. (1998). *Moving up the Organization in Facilities Management: Proven Strategies to Increase Productivity in your Workforce*. Scitech Publishing.

Green, G. M. and Baker, F. (1991). *Work, Health, and Productivity*. Oxford University Press.

Hamer, J. M. (1988). *Facility Management Systems*. Van Nostrand Reinhold.

Hesselbein, F., Goldsmith, M. and Beckhard, R. (1997). *The Organization of the Future*. Jossey-Bass.

Kennedy, A. (1998). 'Support Services'. In *Facilities Management: Theory and Practice* (K. Alexander, ed.), E. & F.N. Spon, pp. 134-145.

McGregor, W. and Then, D. (1999). *Facilities Management and the Business of Space*. Arnold.

Moos, R. H. (1996). 'Understanding Environments: The key to improving social processes and program outcomes'. *American Journal of Community Psychology*, 24, pp. 193-201.

Park, A. (1994). *Facilities Management: An Explanation*. Macmillan.

Rondeau, E. P., Brown, R. K. and Lapides, P. D. (1995). *Facility Management*. John Wiley & Sons.

Schulze, P. C. (1999). *Measures of Environmental Performance and Ecosystem Condition*. National Academy Press, Washington.

Tompkins, J. A. (1996). *Facilities Planning* (2nd edition). John Wiley & Sons.

van Delinder, T. (1997). 'Creating a Progressive Office Environment'. *Facilities Management Journal*, May/June, pp. 18-22.

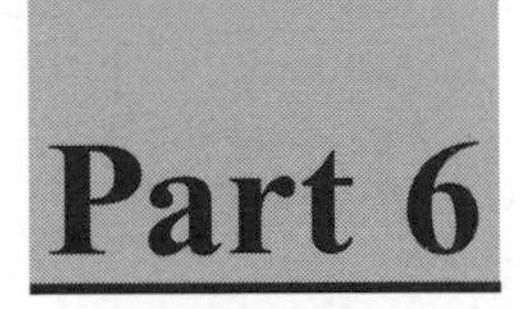

Part 6

Financial management

As stated earlier, facility management is about improving quality, reducing cost and minimizing risk. Therefore it is not surprising that effective financial management is a key activity for the facility manager. Increasingly the challenge is to do more with less, to look for smarter solutions and to minimize resource usage. New ideas are evaluated on a cost and benefit basis, often over a time horizon of many years, to determine the most cost-effective option. This involves assessing financial return together with more qualitative criteria such as image, quality, flexibility and environmental impact.

The construction of new facilities often is the most significant single capital expenditure that an organization will pursue in a short period of time. It is obviously important to get value from endeavours of this sort. Feasibility study is the generic name given to financial evaluations that are performed prior to making a decision to proceed with construction. It is essentially an investment tool, and if used correctly will identify the option that represents the greatest overall return. Implicit in the process is that there are at least two options being evaluated (sometimes one can be 'doing nothing') and that the identified return is sufficiently balanced with the associated risk to make it a worthwhile project.

Discounted cash flow is frequently employed to help make decisions between competing investment options that span more than a couple of years. While the concept of discounting may appear controversial to some (and confusing to others), its basis cannot be challenged, although its application occasionally can. Decision tools, such as net present value, benefit-to-cost ratio and internal rate of return, are commonly used to rank alternatives and determine the recommended outcome. Risk analysis is critical in arriving at a correct decision.

Once a decision has been made to proceed with new construction, a range of budgetary processes is required. Budgets are necessary not only to ensure that available funds are wisely distributed throughout all elements of the project, but that targets are set against which actual performance can be compared. Without budgets it is difficult to manage projects of this nature and the likelihood of significant financial loss is increased.

Budgets extend to other activities than new construction. Operational budgets must allocate annual resources to all aspects of the facility, including cleaning, energy, maintenance, replacement, insurances, security, support services, technology infrastructure, contingencies, etc. Budgets must be agreed with senior management and controlled to ensure that agreed commitments are not exceeded. Forecasts of future cash flows may also be useful.

From an accounting perspective, the financial management of facilities also includes tax implications. Organizations commonly pay tax on income earned and can claim deductions for all costs incurred in its generation. This includes operational costs that are revenue-based and not of a capital improvement nature. However, the latter can be depreciated over a period of time and therefore also offset income tax payments. Making effective decisions that minimize tax deductibility and depreciation in comparison to gross income will realize benefits to an organization's bottom line.

It is generally regarded that cost cannot be controlled, but it can be managed. Perhaps a simple play on words, but it indicates an important concept that cost overruns (and under-runs) will always occur, but what is more relevant is that the overall expenditure is within expected limits. This implies that cost variances are managed so that unavoidable expenses in one area are recovered from another. Predictability in the whole process is the key, with an aim to ensuring that the need to ask for more funds towards the end of the financial year is avoided.

Financial management occurs throughout all phases of a facility's life cycle. Feasibility studies are typically at the concept stage, although can reoccur in smaller instances at any time. Budgets relate to occasional new construction projects and to annual operating activities. Tax depreciation looks into the future to predict the decline in value of assets and proportion them across the corresponding time frame. Even at the end of a facility's life there are demolition or disposal costs that must be considered.

This part examines the important area of financial management and the tools that are available to assist. Most facility managers would employ specialist advice on major projects, but are normally responsible for setting and monitoring their own budgets each year. Chapter 16 discusses the types of feasibility studies that are common and the methods used to arrive at correct investment decisions. Chapter 17 explores budgets and budgetary control including the contemporary issue of dealing with budget cuts. Chapter 18 provides some general background on tax depreciation as it applies to built assets.

Facility managers need to have well-developed financial skills in addition to requisite management and communication attributes. They must be able to seek out new ways of doing things at equal or higher quality but lower cost, and continually look for opportunities to add value and support core business tasks.

16 Feasibility studies

16.1 Introduction

Feasibility studies are concerned with any major decision to invest money in new initiatives. These initiatives may relate to facilities, such as a new building or the refurbishment of an existing building, or other business activities, such as the manufacturing and marketing of a new product or the creation of alliances with external organizations. Nevertheless, a feasibility study (or development feasibility analysis) is often associated with property transactions to determine financial return and payback period for investment opportunities.

Facility managers must understand the mechanics of a feasibility study and how to interpret the results. Even where the responsibility of the study is given to an external consultant, the outcome is merely the basis for a decision by the organization within its particular context. The same feasibility study might be attractive to one organization and rejected by another, despite the details of the project being identical. Feasibility studies concentrate on measuring financial performance, although they may describe other less tangible issues, and are usually summarized as ratios of return and yield. Risk management must be a key component of the investigation.

There are many types of feasibility studies, but they are all characterized by their methodical assessment of benefits (inflows) and costs (outflows). Benefits are calculated as the dollar worth of the development project upon completion and becoming fully operational as a 'going concern'. Costs include all the expenditure that is required to bring the development to completion. Finance and holding charges must be considered. The bottom line is usually referred to as the developer's profit or margin.

16.2 The developer's equation

The developer's equation takes the price-determining input factors into account, together with the likely returns from the completed project. For property transactions, input factors may include the cost of land, building construction and finance (interest) charges. The equation can be simply expressed as:

Value = Land + Building + Finance + Profit

Therefore, the developer's profit is calculated by transformation of the equation as follows:

Profit = Value – Land – Building – Finance

This equation (in either form) underlies the basic methodology of a feasibility study. The aim is to maximize profit by delivering a project with high market value while keeping development costs as low as possible. Projects are feasible if they generate benefits in excess of costs, and the resultant profit is more than what would be realized via other investment opportunities. However, high profit may also entail high risk, so a balanced decision is necessary to ensure that the project matches the attributes of the organization and fulfils intended requirements.

Projects that are complex may take many years to develop, and this introduces additional problems of equating money in different time periods so that a proper decision can be made. In cases where projects involve cash flows which span more than a couple of years, the process of discounting is applied to adjust future cash flows into a common base (or present day equivalent). This is not simply an inflation adjustment, but rather one that also takes account of other issues like investment potential and future prosperity.

16.3 Types of feasibility studies

Despite the numerous different versions of feasibility studies, they basically fall into two groups. One is used for short-term investments characterized by profit realization within a couple of years. The investor is more likely to be a developer and the motive is more likely to be making money. The other type is used for long-term investments creating an interest for the investor over two or three years or more. The investor in this case is more likely to be an owner/occupant, such as a government agency or a large corporation. Facility managers are therefore more likely to be involved in the latter type of study. The development period in all cases is the time horizon used for the feasibility study.

16.3.1 Profitability approach

Profitability is the conventional approach employed in a feasibility study for a short-term investment opportunity. It uses the developer's equation to determine the resultant profit which the investment is expected to generate. Value is interpreted as the market value of the completed project. For projects that involve buildings, value is usually calculated using the income approach to property valuation.

The income approach to property valuation determines market value by dividing expected net operating income (NOI) by a factor known as the net capitalization rate. NOI is the annual income that a property generates after all operating costs are deducted. Vacancies and selling costs may form part of this calculation. Once land, building and finance costs are deducted, the developer's profit is determined.

16.3.2 Residual approach

Essentially the residual approach is identical to the profitability approach, and therefore is also used for short-term investments. The difference between approaches is that the bottom line of the analysis is land value. Therefore the developer's equation is transformed as follows:

Land = Value – Building – Finance – Profit

This approach is used to determine if land should be purchased as an investment, or more to the point, at what land price is the project feasible. It is used when the land is not already owned or where the value of the land for sale is to be determined.

16.3.3 *In globo* analysis

Feasibility studies can also be undertaken in respect of land subdivision, and this provides the *in globo* value of the land. It is very similar in concept to the residual approach, and is also a short-term investment method. Value is determined as gross realization of sales, from which is deducted all development costs to arrive at gross *in globo* land value. Land rates, fees and holding charges are removed to give the net *in globo* value.

16.3.4 Discounted cash flow approach

Discounted cash flow (DCF) is used for long-term investment decisions. The DCF approach is quite different from the previous methods because it additionally considers the impact of time value. It is also generally regarded as a superior method for longer-term investments than those that ignore time value (such as simple payback and accounting rate of return methods). Nevertheless, DCF can be used for short-term investments by creating monthly cash flows rather than the typical yearly approach, and applying a monthly discount factor.

In any case, costs are equated with benefits to calculate the net benefit per period, which is then discounted. The sum of the discounted net benefits indicates whether a project is feasible or not. The value of the property is included as a theoretical sale in the final period of the study. It is a more complex method to use, and is more relevant to ongoing facility management than the short-term options commonly used by developers.

16.4 Costs and benefits

The weighing of expected costs and benefits for an investment initiative is one of the major activities in any feasibility study. It is critical that both costs and benefits are considered, otherwise return on the investment cannot be calculated. Ideally benefits should outweigh costs by a sufficient margin to make the investment attractive. However, the assessment of costs and benefits requires a good understanding of the marketplace, not only today but in the future when the investment is operational.

A market analysis is vital to the success of any feasibility study. Its basic objective is to determine that a market exists for whatever the investment is producing, and that positive return is both likely and of significant size to cover development costs. For example, it may be pointless to construct a health club in an area where competition is intense, or it may be unwise to build more office space for rent where existing vacancy rates in the area are high. It is also a question of timing, for today's environment may be quite different to that which may exist in several years' time when the project is finished.

16.4.1 Costs

Most feasibility studies, certainly those concerning real property, deal with costs such as land acquisition, building construction, building operation and finance. But in fact costs can be any initial or recurrent expenditure directly related to the generation of investment return. They represent cash outflows. Where taxation implications are considered, costs can also include income tax, capital gains tax and goods and services tax, against which tax deductions, rebates and depreciation may be offset.

Land costs are made up of the purchase price of the unimproved land together with acquisition costs such as stamp duty, conveyancing fees and selling commissions. The statutory charges applied to land during the development period (such as rates and taxes) should also be included. Land costs can be an input to the calculation of developer's profit, or form the study's bottom line.

Building construction costs are the capital costs of the development other than land acquisition. At the development stage, given that the design is no more than a concept, the construction costs are normally calculated on a cost per square metre basis. Cost modelling using an elemental or functional approach is useful in determining construction cost. The maximization of net rentable space is a primary objective since this will also maximize investment return. Rise and fall in construction costs should be included if applicable.

Operating costs incurred during the development period are considered as an expense incurred in earning income. Operating costs include cleaning, energy, repair, maintenance and other cash outflows, and in special cases may also include occupancy costs (functional-use costs). Where tenants contribute towards the payment of operating costs, the costs are reduced by the amount that is 'recoverable' so that only the developer's portion is considered. For short-term investment projects, operating costs are normally ignored.

The cost of interest payable on borrowings and interest earned on unspent income form part of the feasibility study, unless a DCF approach is used. Interest is treated like a bank overdraft. It is paid when capital is drawn down, and it is received when benefits are accumulated to recover loans. Therefore whenever the account is in deficit interest is paid, and whenever the account is in surplus interest is received. Holding charges are therefore built into the study.

16.4.2 Benefits

Benefits are cash inflows. They may take many forms depending on the type of project being evaluated. For example, benefits may comprise the market value of the property (as indicated by the capitalization of net operating income) or the total sale price of land packages or accommodation units. In the former case, net operating income can be calculated as the rentable floor area multiplied by the rental price per square metre for various classifications of space, minus unrecoverable operating expenses. Benefits may also be derived from the sale of products in the marketplace or the acquisition of a new venture that generates income of some sort. It is usual to deduct 'costs' such as selling costs from gross revenue, thus calculating net revenue (or net return). Selling costs can include advertising, legal expenses, commissions and holding charges (or interest incurred) during the development period.

For rental property or where units of accommodation are being sold, there can be a delay in achieving initial tenants which results in lost income. Furthermore, as tenants leave there can be a delay in finding replacements. An assessment of vacancy rates is therefore necessary to determine the real return, as the theoretical earning capacity of a facility may never be fully achieved. New developments ideally seek pre-commitment in terms of sales or rental agreements to overcome the costs of unrealized income during the selling or letting period. Developers normally make allowance for vacancy rates or must assume that accommodation units will be sold over a period of time. This becomes a key variable in a feasibility study and requires professional market knowledge to get it right.

Feasibility studies involving short-term projects typically have benefits derived from sales, whereas long-term projects typically have benefits derived from rent or lease. Nevertheless in both cases an assessment of the value of the investment is involved. For short-term investments the value of the development forms part of the gross return from which all the costs of development are deducted. For long-term investments a theoretical sale of the property is made in the final year of the cash flow and treated as a benefit. This is called reversion, and recognizes that any project has some value at the end of its 'life'. For real property, this can be quite significant. For some equipment purchases, however, it may be negligible.

16.5 Assessment criteria

16.5.1 Short-term investments

Short-term investments are typically judged on both developer's profit and investment yield. However, risk analysis is necessary to gauge the likelihood of the estimates being realized. The more uncertain the estimate, the higher the profit or yield should be to provide compensation for the risk of loss.

The developer's profit is the amount of money left over after the investment is sold and all development costs have been paid. It is normally expressed as a dollar amount (development margin) and as a percentage (profit) of total development costs. Developer's profit is relevant to short-term investments and must compare favourably with alternative uses of capital.

Investment yield is also used where profit is the basis for decision. Yield is defined as net income generated divided by investment value. For property transactions, value is interpreted as the sum of land, building, finance and profit. Therefore investment yield is simply the ratio of net income to total development costs and can be alternatively described as return on investment.

16.5.2 Long-term investments

Under a DCF approach, developer's profit and investment yield give way to net present value (NPV). Projects are favourable where NPV is positive, and in cases of mutually exclusive investments, the higher the NPV the better. Projects are chosen provided they have sufficient return to cover expected profit and risk allowances. Note that use of a discounted cash flow approach builds in interest paid and received (or debt and equity commitments) as part of the discount rate and therefore interest should not be separately included in yearly cash flows.

NPV is defined as the sum of discounted benefits less the sum of discounted costs over a given time horizon. The further away from the present that costs or benefits are involved, the more they are reduced (discounted) so that their impact is effectively less. This calculation caters for the condition that money has investment potential and is therefore more valuable today than tomorrow. At discount rates of 5% or more, costs and benefits incurred or received after 20–25 years are rendered negligible.

Internal rate of return (IRR) is another useful assessment criterion. IRR determines the level of profitability and risk contingency implicit in a project. At a given real discount rate (based on the true time value of money), the difference between the IRR and the discount rate can be considered as equal to the profit and risk allowance. The larger the difference the more attractive is the investment.

Investment ratios such as benefit-cost ratio and saving-to-investment ratio can also be used to help judge return. They generally apply to discounted cash flow applications, but some ratios can also be used in non-DCF applications.

A DCF approach by its nature involves a comparison between two or more alternative investments. Most feasibility studies involve some form of comparison, but where no explicit alternative is identified, an implicit comparison still occurs. The discount rate reflects the opportunity cost of investing in the financial marketplace at a secure rate of interest and thus equates to a theoretical 'do nothing' option.

16.5.3 Risk assessment

All feasibility studies involve prediction of future events and therefore uncertainty. Risk analysis should be employed to assess the impact of influential variables on the decision criterion. Sensitivity analysis is the most common form of risk analysis particularly for short-term investments. Failure to properly consider risk exposure is equivalent to professional negligence.

For short-term investments, influential variables often concern the calculation of gross return. This may involve rent levels, vacancy rates, selling prices, letting/selling periods and the selection of capitalization rate. For long-term investments, discount rate selection and study period additionally become significant. Reversion in DCF is discounted and therefore is generally not a key factor.

16.6 Conclusion

Feasibility studies are an important part of making effective investment decisions. While they focus on profit issues, often to the complete exclusion of subjective, social or environmental concerns, they nevertheless identify investments that yield positive returns and contribute to the prosperity of stakeholders. A DCF approach is generally regarded as a superior method for all but the most short-term projects. Both market analysis and risk assessment are vital to a successful study.

In the end a feasibility study is merely a prediction of a future outcome. Variations will occur, so it is important that expected return is sufficient to cover the risk involved. For example, investing $1 million to make $1 extra is ridiculous. So not only does the study need to identify which projects are likely to deliver positive returns, but also the magnitude of these returns is important in the context of the organization and its other activities.

References and bibliography

Alexander, K. (1996). *Facilities Management: Theory and Practice*. E. & F.N. Spon.

Atkin, B. and Brooks, A. (2000). *Total Facilities Management*. Blackwell Science.

Barrett, G. V. and Blair, J. P. (1982). *How to Conduct and Analyze Real Estate Market and Feasibility Studies*. Van Nostrand Reinhold.

Barrett, P. (1995). *Facilities Management: Towards Best Practice*. Blackwell Science.

Clifton, D. S. and Fyffe, D. E. (1977). *Project Feasibility and Analysis: A Guide for Profitable New Ventures*. John Wiley & Sons.

Cotts, D. G. (1999). *The Facility Management Handbook* (2nd edition). AMACOM.

Gilbert, B. and Yates, A. (1989). *The Appraisal of Capital Investment in Property*. Surveyor Publications.

JLW Research and Consultancy (1992). *Capitalisation and Discounted Cash Flow Valuation: Bridging the Gap*. JLW.

McGregor, W. and Then, D. (1999). *Facilities Management and the Business of Space*. Arnold.

Park, A. (1994). *Facilities Management: An Explanation*. Macmillan.

Robinson, J. R. W. (1989). *Property Valuation and Investment Analysis: A Cash Flow Approach*. Law Book Company, Sydney.

Roddewig, R. J. and Shlaes, J. (1983). *Analyzing the Economic Feasibility of a Development Project: A Guide for Planners*. American Planning Association.
Rondeau, E. P., Brown, R. K. and Lapides, P. D. (1995). *Facility Management*. John Wiley & Sons.
Stevens, D. (1997). *Strategic Thinking: Success Secrets of Big Business Projects*. McGraw-Hill.
Tompkins, J. A. (1996). *Facilities Planning* (2nd edition). John Wiley & Sons.
Wright, M. G. (1990). *Using Discounted Cash Flow in Investment Appraisal*. McGraw-Hill.

Budgetary control

17.1 Introduction

Since the 1980s facility managers have had to operate in an environment of heightened business competition and budget cuts. Senior management often seek to reduce the costs of doing business when faced with increased external competition, and facilities are seen as one area of expense where savings must be found. As a result, facility managers over the years have got used to the idea of 'doing more with less'. However, effective financial management is necessary to avoid the situation becoming one of 'doing less with less'.

Budget 'cuts' are the result of senior management's attempts to keep their businesses afloat. This restricts the facility manager's work and limits important goals like planned maintenance, job security for in-house staff and providing them with appropriate development and training opportunities. Keeping operational costs down often result in a focus on energy and maintenance, yet usually to make long-term savings in these areas additional money first needs to be spent. The facility manager's role can become more operational as a direct result of trying to keep everything going under enormous budgetary pressure.

To avoid the vicious circle of budget cuts each year, facility managers need to think more strategically. In order to do more with less, a completely different approach may be necessary. A quantum shift in the nature of facilities that are provided and how they are used to support business may be needed. Improvements in productivity and business profit are the key, so the focus is moved from one of cost reduction to one of innovation and process re-engineering. The new goal can become 'spending more to make more'. For this to succeed, the facility manager needs to play a much more influential role in overall business decision-making.

17.2 The budgeting process

Budgeting is the disciplined pre-determination of cost, and as such is a significant element of planning and control for any business or project. It portrays the likely consequences of design, management or operational decisions in terms of dollars. It also identifies major cost factors and areas of risk. It is the first step in the overall planning and control process for any activity.

Accountants readily recognize the importance of budgeting in business concerns. In these cases the budget provides management with the right tool to make effective use of the capital at its command. The basic budget is seen by management as both an early warning system and a frame of reference for evaluating the financial consequences of operational decisions. The degree of sophistication in a company's budgeting system will usually depend on its size, but even small businesses will find that an elementary level of budgeting will assist in the control of the business.

The rising costs of facilities in recent years have emphasized the need for careful control of cost. The budget is the mechanism by which control can be achieved. It provides a standard for comparison, without which there can be no benchmark of financial performance.

The budget becomes the mechanism within which the facility management team must work to achieve their objectives. It provides a considered measure for comparison and judgement of actual results. Thus it is useful in controlling all aspects of cost in both the present and also future years. However, budgets should be realistic. The setting of unrealistically low budgets may initiate a large amount of investigation into new methods or ways of doing things, but where this is too difficult may in the end lead to an over-run of cost or a reduction in quality. Setting excessively low budgets to force change can be counter-productive.

The process of budgeting should be understood within the organization at all levels. An ideal arrangement is that the process should support the following principles:

1. Transparency of information.
2. Dissemination of divisional budgets throughout the organization.
3. Consistency of budgeting process from one year to the next.
4. Accountability of performance.
5. Safety net mechanisms for divisions when things go wrong.
6. Reward for efficiency.
7. Equity.
8. Reward for entrepreneurialism or innovation.
9. Collective decision-making.

A culture needs to be developed where allocated funding does not have to be spent or lost, but that divisions can save for future years to cover major items of expenditure. This is particularly important for the facility management division.

17.3 Preparing a budget

Budgets are allocated by senior management. However, usually there is an opportunity for each business unit to prepare their own budget for approval. If the facility management group is treated like a business unit, then the facility manager would be required each year to prepare a forecast of what funding is necessary.

Budgeting can be a strategic process where the amount asked is in excess of what is really needed, so that when senior management reduce allocations across the board, the division actually gets enough to manage its affairs properly (van Haasteren Jr., 1996). But in most cases budgets are largely based on plus or minus adjustments to what was allocated in the previous year. Rather than have budgets progressively reduced or just keeping pace with inflation, facility managers should proactively look at ways to improve business performance and ask for extra money to achieve it.

Budgets must be clearly set out and professionally presented. There is no universal format, but it is useful to include last year's performance, this year's request and next year's plan to place the numbers in context. Budgets should be fully explained and justified by proper analysis. The use of illustrations such as pie charts and trend lines can significantly improve understanding by management (Kozlowski, 1999).

Facility management covers a wide range of support services. The budget should categorize the major areas of activity using cost centres. Using this approach it can be clearly seen where the money is actually going, such as 20% on maintenance, 10% on energy, 5% on furniture, 10% on security, etc.

Comparing past performance with industry benchmarks helps to provide the necessary justification for budget allowances. Business performance indicators can be used to demonstrate that the funds have been wisely used to bring gain to the organization, to increase flexibility, minimize risk exposure, improve quality and raise customer satisfaction levels.

It is normal for a detailed budget to be prepared and submitted to senior management for decision. It is also normal that divisions receive a single line allocation, equal to or less than the amount requested. Following allocation, divisions will rework their budgets to manage responsibilities within the funding provided to them. Large reductions in allocated funds will engender significantly more rework at the divisional level.

17.4 Historical trends

Trends depicting past performance are an important part of budgetary control. Not only does past information provide an effective basis for predicting future information, but it also provides a context for allocation decisions. For example, costs of Internet access for the organization might be rising at an exponential rate, or electricity costs might be slowly falling over a number of years. Illustrating these trends graphically is the best way to communicate facility impacts. Presentational software tools enable impressive documents to be easily prepared.

Forward planning is a critical aspect of prudent financial control. Judgements about the future are drawn from trends of what has gone before. While events of significant change can and do occur, there is an underlying assumption that the future will generally be consistent with the past.

17.5 Dealing with budget cuts

Budget cuts are usually not one-off events but rather lead to a permanent reduction in resources for future years as well. Progressive budget cuts can be even more devastating. Business downsizing will impact on total facility costs (by the reduction in required infrastructure) and on unit costs (by the need to become more efficient per m^2 of space or per number of employees). The impact of budget cuts can be far-reaching within the organization and within management divisions.

There are clearly limits to how efficient organizations can become, and many organizations are already close to a practical optimum. To find further efficiency may require significant re-engineering of the ways in which things are done, and dealing with associated staffing implications. Yet cutting costs is part of the overall process of managing within tight financial constraints, and inevitably a balance needs to be found between cost and quality.

It is increasingly important for some aspects of facility operation to be made self-funding. This could apply to support services like food outlets, childcare, health and fitness, and so on. Under-utilized facilities are a clear target for change. In some cases facilities can be made available to other organizations for a fee in order to generate a valuable income stream.

In other cases, outsourcing may be relevant to reduce the size (and cost) of the workforce. However, lower costs are not always the actual outcome, and issues like increased expertise and decreased turnaround time are important considerations.

Budget cuts require a proactive plan of attack. With improvements in technology continually becoming available, new solutions to old problems can always be found. An analysis of their feasibility is required, but where total gain is anticipated the proposals should form part of future budget requests.

Once a budget is allocated it is normal for it to become an immovable financial constraint. Yet in many organizations new contracts, clients, unanticipated success or failure will result in changes to budget allocations at various times in the year. The budgeting process can be very fluid and under continual review. However, a lack of permanency in the allocation will hamper the attempts of divisions to plan and to find innovative solutions to reduce costs or increase productivity.

Dealing with budget cuts is a difficult process. 'Textbook discussions of maintenance budgeting reveal little of the trauma that can result from budget cuts [and . . .] dry recitations about cost estimating and resource allocation don't do much good when the budget proposal you've slaved over lies in tatters as a result of newly announced spending reductions' (Hounsell, 1997). The larger the facility management operation, the more complex the process becomes.

17.6 Cost monitoring and financial control

Budgets are targets that guide and help control activity costs. While the overall budget may be considered as a maximum limit, the distribution of funds within the management division can be considerably more flexible. Therefore an unexpected over-run in one area can be funded by finding a saving in another area, so the net result is no change.

Actual performance therefore has to be closely monitored and continually compared to budget forecasts. In this way the budget acts as a control document. Differences between actual and planned expenditure require review and perhaps corrective action. Monitoring also enables future budgets to be improved.

Monthly comparisons of performance are vital. Performance indicators can assist in highlighting those areas that are going well and those that are not. The facility manager needs to continually act on this information and fine-tune operations so that final targets are achieved. If no action is taken, small problems can turn into big ones and the potential to recover is diminished.

Senior management may also be interested in monthly comparisons of performance. Where over-runs are incurred, plans for redressing the situation may require formal submission and endorsement. In some organizations, cost information is online for all managers to access, giving an up-to-date picture of financial performance.

17.7 Cost centres

Cost centres are accounting categories used to help keep track of expenditure and to manage budgets effectively. Cost centres will vary from one organization to another, but need to reflect core activities. Some typical facility management cost centre examples for office-based facilities include:

Salaries
1. Management.
2. Maintenance.
3. Information technology.
4. Security.
5. Food preparation.
6. Staff development and training.
7. Consultants.

Space costs
1. Rent.
2. Rates and charges.
3. Insurance.
4. Relocation.

Energy
1. Electricity.
2. Fuel.
3. Water.

Maintenance
1. Cleaning.
2. Repair.
3. Replacement.
4. Vandalism.
5. Emergencies.
6. Groundworks.
7. Testing.

Information technology
1. Hardware.
2. Software.
3. Networking and Internet costs.

4. Data storage and archive.
5. Presentations and special needs.

Furniture

1. New furniture.
2. Repairs.
3. Satellite workplaces.
4. Churn.

Security

1. Patrols.
2. Access controls and equipment.

Other

1. Capital works.
2. Contingency.

The list is not intended to be exhaustive nor suggestive of what should or shouldn't be included within the facility management division. Capital works cost centres would be separately subdivided by project and have their own detailed budgets and monitoring procedures.

17.8 Contingencies

Budgets are forecasts of future events and therefore involve uncertainty and risk. Contingencies are a common mechanism to allow for the risk of unexpected events. They are often a percentage allowance used to pay for a range of miscellaneous items not specifically identified elsewhere.

Contingencies need to be appropriately managed so that they are not totally expended in the first few weeks of the year, but progressively drawn on as the need arises. Contingencies are not intended to cover budget over-runs in the various cost centres, although at the end of the year they might be used to help balance the overall budget. They are more commonly used to deal with emergencies and disaster recovery needs, or special events or opportunities that arise after the budget allocation process has been completed.

17.9 Long-term planning

Budgets are typically annual, either based on a fiscal or calendar year. Facilities like buildings are long-term assets and their needs do not necessarily fall neatly within annual time horizons. For example, extensive façade cleaning might occur every five years or HVAC equipment replacements every twenty years. Therefore in addition to annual budgetary control there needs to be a system of planning for these irregular but significant cost items.

Long-term asset planning is undertaken in conjunction with life-cost planning and analysis techniques that adopt a whole-of-life approach. Such techniques require a significant amount of skill and construction knowledge and may involve the engagement of specialist consultants.

Cash flows are usually prepared based on estimates of future costs and their likely timing. While this process does not mean that such work must take place at the designated time, whether necessary or not, it does enable proper consideration and understanding of the real costs of owning and operating an expensive asset. Major refurbishments and upgrades to services, equipment, software, etc., must be planned well in advance so that appropriate funds are available when needed.

17.10 Sinking funds

Sinking funds are usually created to pay for the replacement of capital assets at some time in the future. They work on the concept that money is invested in an interest-bearing account so that it can grow to the amount needed to pay for the replacement in the future. In many areas of business activity, sinking funds have now given way to leasing arrangements where the equipment (in this case) is never owned but paid for as monthly contributions to the leasing agent.

Capital assets are eligible for tax depreciation allowances. These are also a way of allowing for future replacement, and in some cases may be allocated to a sinking fund cost centre for such purpose. Organizations need to manage these key assets so that financial problems do not occur later in their life. Discounting techniques are used to calculate annual sinking fund contributions and the process is often left in the hands of the corporate accountant. Nevertheless, the facility manager has access to these funds to support and maintain facilities for as long as they are needed, and certainly needs to be aware of what funding allocations are set aside.

17.11 Major capital works

Organizations plan to grow and expect to undergo continual change as they adapt to new environments and new opportunities. Therefore facility costs are not confined to keeping existing infrastructure operating, but also to procuring new infrastructure. New infrastructure involves significant capital investment and in many cases may include design processes and external consultants.

Budgeting for capital works is almost a completely separate process to operational activities. In many cases funding comes from separate sources. This can impact on design philosophy insofar as minimizing capital expenditure to comply with funding constraints may deliver projects that have high operating costs in their future. Contemporary thinking is that capital and operating budgets should be linked so that an opportunity is created to spend more initially to realize long-term savings.

Capital works decisions are made on the basis of some form of feasibility study, and take into account a range of issues, not purely monetary ones. Nevertheless, the facility manager is at the heart of the procurement process and usually has a major influence in asset acquisition policy and direction.

17.12 Construction budgets

Not only is a budget a control mechanism for business, it also has an important use at an individual project level, such as the construction of a new building. A budget prepared for a construction project gives the client and design team the means to plan, co-ordinate and control the project and becomes a system for evaluating the actualities of design decisions. The budget establishes cost limits for key areas or elements of the building. While the total project may be within budget, examination of these key areas may show major discrepancies.

These days construction cost (if not life-cost) is usually considered as a design limitation. Without budgets the limiting factors of design are function and aesthetics, and cost becomes merely the result of decisions made about these limiting factors. The budget more correctly should set a limit within which the design team can work, for in its absence the designers are not aware of the true scope of the project or what the client can afford. The cost of a building is determined by every decision made, and just as initial design decisions will affect a continuing series of interrelated developments throughout the design period, decisions on costs will also affect and be affected by the development of design ideas.

The budgeting process relates design development to costs. It gives the designer the means to forecast the probable outcome of design decisions. The budget also allows for investigation of alternative design solutions while still keeping within the existing budget. An 'open purse' attitude will inevitably result in an over-design, which may functionally fit the client's criteria but be unnecessarily expensive. In this respect the budget can ensure that value for money is a continuing objective.

References and bibliography

Alexander, K. (1996). *Facilities Management: Theory and Practice*. E. & F.N. Spon.

Ashworth, A. (1997). *Cost Studies of Buildings* (2nd edition). Longman.

Atkin, B. and Brooks, A. (2000). *Total Facilities Management*. Blackwell Science.

Barrett, P. (1995). *Facilities Management: Towards Best Practice*. Blackwell Science.

Bernard Williams Associates (1999). *Facilities Economics*. Building Economics Bureau Limited.

Cotts, D. G. (1999). *The Facility Management Handbook* (2nd edition). AMACOM.

Damiani, A. S. (1998). *Moving up the Organization in Facilities Management: Proven Strategies to Increase Productivity in your Workforce*. Scitech Publishing.

Hounsell, D. (1997). 'Surviving Budget Cuts: The ax has fallen. Now what? Facilities offer strategies for thriving in the aftermath'. *Maintenance Solutions*. <http://www.facilitiesnet.com/fn/NS/NS3m7ka.html>

Kozlowski, D. (1999). 'Capital Ideas: The Illustrated Budget'. *EducationFM*. <http://www.facilitiesnet.com/fn/NS/NS3e9bh.html>

McGregor, W. and Then, D. (1999). *Facilities Management and the Business of Space*. Arnold.

Moos, R. H. (1996). 'Understanding Environments: The key to improving social processes and program outcomes'. *American Journal of Community Psychology*, 24, pp. 193-201.

Park, A. (1994). *Facilities Management: An Explanation*. Macmillan.

Rondeau, E. P., Brown, R. K. and Lapides, P. D. (1995). *Facility Management*. John Wiley & Sons.

Stevens, D. (1997). *Strategic Thinking: Success Secrets of Big Business Projects*. McGraw-Hill.

Tompkins, J. A. (1996). *Facilities Planning* (2nd edition). John Wiley & Sons.

van Haasteren Jr., G. (1996). 'Budgets and Beyond: Winning the money game requires knowing the rules and the players'. *Maintenance Solutions*. <http://www.facilitiesnet.com/fn/NS/NS3m6lh.html>

Tax depreciation

Peter Smith

18.1 Introduction

Taxation allowances for the depreciating value of property enable owners to reduce their assessable income and, hence, increase their returns on the property. Often referred to as 'tax depreciation', these allowances can have a significant impact on financial returns from property ownership. It is not unusual for tax savings of millions of dollars on high value transactions to be achieved, but even on smaller projects tax depreciation can make the difference between a viable proposition and a non-starter.

These allowances are set by government legislation and, accordingly, can vary widely from country to country. Where these allowances apply, tax depreciation analysis should be an integral component of project feasibility studies. Accounting for property depreciation will normally be required throughout the effective life of the property.

18.2 Tax depreciation defined

Depreciation can be defined as a decrease in the value of an asset over its economic life due to factors such as general wear and tear, ageing, deterioration and obsolescence. However, it does not relate to sudden destruction or damage resulting from fires, accidents, storms or disasters. Depreciation applies both to tangible property such as machinery and buildings and to intangibles of limited life such as leaseholds and copyrights. A prime example is the purchase and use of a new motor vehicle for business purposes. The vehicle is a business asset, which begins to lose value as soon as it is driven out of the showroom and continues to lose value over its years of operation until the vehicle is of little or no value to the business. Measuring the loss in value of this asset is known as depreciation.

The motor vehicle analogy is equally applicable to built facilities and the plant and equipment housed within these facilities. Whilst most property investors expect their property to increase in value over time, this usually occurs due to the increase in the land value of the property and other 'non-building' factors. However, the actual building and its various plant and equipment actually deteriorate in value from the point of completion or installation.

Tax depreciation refers to tax allowances made by governments, which compensate property owners for this deterioration. It usually only relates to income-producing assets used for business purposes. In this context, depreciation is considered as a business expense for which owners are compensated. It originated in the desire of governments to encourage capital expenditure in manufacturing. Its scope has been broadened over the years to encompass most types of capital expenditure on real estate although the type of expenditure for which tax depreciation is available varies widely by country.

Tax depreciation allowances enhance an owner's return on their property by allowing them to reduce their level of income tax. In other words, depreciation is claimed as a tax deduction and, hence, can lead to improvements in an owner's 'after-tax' returns on a property. Typical properties where depreciation may be applied include buildings, furniture, equipment and machinery.

18.3 Types of tax depreciable property

Property is either tangible or intangible. There are two main types of tangible property:

1. *Real property.* Real property is land, buildings and generally anything built or constructed on land, growing on land or attached to the land.
2. *Personal property (plant).* Plant includes articles, machinery and tools. Examples include computers, electrical tools, furniture and fittings, furnishings, carpet and curtains, manufacturing machinery and motor vehicles.

Intangible property is generally any property that has value but can't be seen or touched. It includes items such as computer software, copyrights, franchises, patents, trademarks and trade names.

Tax depreciation allowances can apply to both tangible and intangible property depending on the respective taxation legislation of a country. With respect to real estate, land is the main component not considered to be a depreciable asset. Even though land values can fluctuate quite markedly (up and down), land is not considered to be a depreciable asset as it does not wear out like buildings or equipment.

The main components that may be eligible for depreciation are, firstly, capital expenditure on the construction or refurbishment of a building and, secondly, plant and equipment used within a building. The legislative definitions of depreciable plant and equipment vary widely but typically include items such as carpet, air conditioning, electrical fittings, furniture, equipment and machinery. Each country has different rules regarding who, if anyone, can claim such allowances. For example, in the United Kingdom, a property developer who intends to sell a property on completion of construction cannot claim capital allowances as they are using the building as trading stock. However, if the developer intends to manage and continue to own the property they may be entitled to claim allowances.

18.4 Significance

The significance of tax depreciation allowances for property investors will depend on the legislative taxation provisions in-country. Often this legislation is complex and requires the use of expert property taxation consultants to enable owners to correctly claim and maximize depreciation deductions and maximize their after-tax return from their investment. In many cases, depreciation allowances are provided for not only new capital building work, but also renovation and alteration work on existing buildings. Many owners are unaware of their full entitlements and/or receive poor advice and, hence, do not maximize their potential returns.

The following demonstrates the significance of depreciation allowances in Australia. Initial construction costs can be depreciated at between 2.5–4% per annum depending on the type of building and the year of construction. However, plant and equipment can be depreciated at much higher rates ranging from 5–50% per annum and, in some cases, 100%. Table 18.1 shows the result of a detailed analysis carried out by Napier and Blakeley (2001) of the typical plant content in a wide range of building types. This plant content is expressed as a percentage of overall capital expenditure on the building excluding the land component.

Table 18.1 Average depreciable plant content of income-producing properties (Australia)

Type of property	Average plant content
Prestige CBD office block	35–45%
Older style CBD office block	25–35%
Suburban low rise office block	20–30%
Full office refurbishment	40–80%
Office fit-out	50–90%
Computer centres	30–50%
Regional shopping centre	25–35%
Suburban shopping centre	15–25%
5 star hotel (including fit-out)	45–55%
3 star hotel (including fit-out)	35–45%
Quality strata residential units	20–30%
High-tech industrial buildings	25–30%
Industrial (high office content)	20–25%
Industrial (low office content)	10–20%
Private hospitals/nursing homes	30–40%
Sports centres	15–30%
Car parks	5–15%

Source: *Napier and Blakeley (2001)*

Depreciation allowances clearly provide owners with the opportunity to reduce the operational costs associated with their investment and, hence, can be a very important part of the overall facility economics approach to asset management.

18.5 Calculating depreciation

The calculation of depreciation varies from country to country. Generally, depreciation is based on historical cost rather than on probable replacement cost, but many countries have complex legislation in place relating to the calculation of the depreciable value of assets. Nevertheless, some common approaches to the measurement of depreciation exist.

The first requirement is normally to establish the initial cost of the asset. The second requirement is to establish the time period over which the assets will retain some value in a business sense. This is often referred to as the useful life of the asset after which it will have worn out and lost all value. Thus, depreciation is usually calculated on the basis of acquisition price and the useful life of the asset.

Most assets have a useful life greater than one year and thus depreciation costs are normally spread over a number of years. The most common accounting systems used to calculate these costs are the prime cost (straight line) and the diminishing value methods of depreciation.

18.5.1 Prime cost method

The prime cost or straight-line method is generally the most common accounting method for depreciating assets. An annual deduction is calculated by multiplying the depreciation rate for that item by the initial cost of the item. The depreciation rate is a percentage based on the effective life of the item. The data required is therefore the initial cost of the asset and its estimated useful life.

The following is a simple example of how this method is applied for an asset costing $10 000 with an effective life of five years and a depreciation rate of 20% per annum.

Income year (opening)	Undeducted cost deduction $	Depreciation (closing) $	Undeducted cost $
Year 1	10 000	2 000	8 000
Year 2	8 000	2 000	6 000
Year 3	6 000	2 000	4 000
Year 4	4 000	2 000	2 000
Year 5	2 000	2 000	0

The depreciation deduction is spread out evenly over time.

18.5.2 Diminishing value method

Using the diminishing value method, the depreciation rate is generally one-and-a-half times the prime cost depreciation rate for that item. The deduction in the first year is calculated by multiplying the depreciation rate for that item by its initial cost. In subsequent years, the rate is applied to the amount remaining after deducting the depreciation claimed in all previous years from the initial cost.

The following is a simple example of how this method is applied using the same example as above. The prime cost depreciation rate is multiplied by 1.5, which equates to 30%.

Income year (opening)	Undeducted cost deduction $	Depreciation (closing) $	Undeducted cost $
Year 1	10 000	3 000	7 000
Year 2	7 000	2 100	4 900
Year 3	4 900	1 470	3 430
Year 4	3 430	1 029	2 401
Year 5	2 401	720	1 681

Under this method, the depreciation deduction is higher in the early years but declines over time. The asset will continue to be written off over further years or until it is sold or replaced.

18.6 International comparisons

This section provides a general summary of how a randomly selected group of countries provide (or don't provide) for the depreciation of assets in the construction/property sector. They provide a good example of the varied range of approaches used around the globe. This information was obtained from the website of Price Waterhouse Coopers (2001) and was current as at January 2001.

It must be remembered, however, that taxation is a complex field impinging on many areas. Other taxation allowances (rental income, capital gains tax, income tax levels, corporate tax levels, goods and services tax, and the like) all affect the manner in which the depreciation of assets may be calculated.

18.6.1 United Kingdom

Depreciation is generally not deductible, but capital allowances may be claimed on certain plant and machinery in commercial buildings at a rate of 25% per annum on a declining balance basis. A reduced rate of 6% per annum applies to certain assets that have an expected useful life of at least 25 years, but this will not usually apply unless the building is industrial. Capital allowances can also be claimed for 100% of the original cost of certain commercial buildings in enterprise zones, and 4% per annum on hotels and many industrial buildings located elsewhere in the UK. Most expenses, other than those of a capital nature, can be deducted. The construction, alteration or refurbishment of any commercial building, and the freehold sale of commercial buildings less than three years old, are liable to a standard VAT rate of 17.5%. The sale of older commercial buildings, and the grant of leases for any commercial building, are generally exempt.

18.6.2 United States

Construction and acquisition costs are allowed to be depreciated at various rates as a deduction against federal tax liability. Generally, entire commercial construction and acquisition costs are depreciated over a 39-year period, and residential properties over 27.5 years. Land improvements or other components of the property may be depreciable over a shorter period of time but costs attributable to land acquisition are not depreciable. Cost segregation studies can be used to identify costs that qualify for federal tax lives of five and seven years for personal property and 15 years for land improvement property. The term 'cost segregation' describes the process of splitting the acquisition cost or construction cost so that individual assets are allocated to their appropriate tax life. A case example of this system is provided later.

18.6.3 Australia

Deductions against rental income may be available for depreciation and building capital allowances. Certain components of a building (new or second-hand) may be considered to be plant and equipment. These are depreciable for tax purposes generally over their useful lives. Examples of qualifying items include air conditioning, certain electrical installations, lift equipment and fixtures and fittings. Values for depreciation depend on an allocation of the purchase price of the property that may be specified in purchase contracts or determined by valuers. There is no definition of plant under the relevant taxation legislation, however, one income tax ruling details over 1500 items of plant that may be depreciable. Depending on the property's specifications, fit-out and sale contract conditions, owners may be eligible to claim between 20–40% of the building's value as depreciable plant. The building (or structural) part of a property may be eligible for a capital allowance write-off. For buildings constructed prior to 20 July 1982 no write-off applies. For buildings constructed after this date the write-off rate is either 2.5% or 4% p.a. The capital allowance (unlike depreciation) is always calculated on the original construction cost. The costs of improvements or extensions may also qualify. Interest on borrowings used to acquire property is deductible against rental income. Other property costs incurred in deriving rental income such as insurance, repairs and maintenance and property management are also deductible. Costs which are capital in nature, such as stamp duty and legal costs incurred in relation to the acquisition of property are not deductible but form part of the capital gains cost base. A goods and services tax (GST) of 10% was introduced in 2000. This generally applies to the sale of newly constructed buildings but not the sale of buildings completed prior to the implementation of the GST.

18.6.4 Hong Kong

No tax depreciation allowance on a building or property is available to an individual investor who is liable to pay property tax. However, corporate investors are entitled to tax depreciation allowances on their property. Certain components of a building (new or second-hand) may be considered to be part of the physical plant and equipment. These are depreciable by an initial allowance of 60% on the acquisition cost in the year of acquisition and an annual depreciation allowance thereafter of 10–30% on the depreciated value, depending on the nature of the plant and machinery. For example, lift equipment, escalators, air conditioning systems, sprinklers and the like are considered to be part of the physical plant and eligible for a 60% initial depreciation allowance and an annual depreciation allowance at the rate of 10% thereafter. The building or structure, or part thereof, other than the physical plant and equipment, may be eligible for a tax depreciation allowance on the costs of construction. If the building or structure is used by the owner, or its tenant, in a qualifying business, such as milling, manufacturing, transportation, public utilities, farming and trade of storage, then an industrial building allowance would be available. An initial depreciation allowance of 20% on the costs of construction is available for the first use of an industrial building, together with an annual depreciation allowance at the rate of 4% on a straight-line basis. For subsequent use of such an industrial building, the annual allowance is computed by referring to the unclaimed tax portion of the residual value based on the unused portion of the building's statutory useful life of 26 years. In respect of buildings or structures other than qualifying buildings, a commercial building depreciation allowance of 4% of the construction costs is available. Corporate investors may also claim a deduction of capital expenditures relating to the renovation or refurbishment of a building or structure, other than a domestic building or structure, over a five-year time frame.

18.6.5 Germany

In Germany, an annual depreciation charge of 2–4% on buildings (exclusive of land) is allowed depending on their dates of completion. Depreciation is calculated on the property's historic cost price. However, under current tax reform proposals the depreciation regime will become less favourable. Rates higher than the aforementioned standard rates may be applied under certain conditions. As from 1999, these higher rates have been replaced by investment subsidies ranging from 10–15% of the cost basis of the building for certain investments. The annual depreciation rates on business fixtures and movable assets depend on each individual asset. VAT applies in Germany but sale and lease transactions are tax exempt in principle. In some cases, developers are allowed to exercise an option with regard to VAT.

18.6.6 New Zealand

A deduction is available for depreciation on the total acquisition cost of the property, excluding land but including any extensions or improvements. Any subsequent revaluation of the property is ignored. The rate of depreciation that can be claimed on a building is generally 4% on a declining balance basis or 3% on a straight-line basis.

Certain components of the building may be considered to be plant and equipment, such as fixtures and fittings, and if so, will attract a higher depreciation rate. Expenses incurred in deriving rental income are generally deductible. Such expenses include interest on loans used to acquire the property, as well as other property costs such as repairs and maintenance, insurance, rates, administration costs and depreciation. Capital expenditures are not deductible, but may be included in the acquisition cost of the property and thus depreciated. A goods and services tax is levied at a rate of 12.5%.

18.6.7 Spain

Generally, an annual 2% (3% for industrial buildings) depreciation charge on property (exclusive of land) is allowed. In certain circumstances these depreciation rates can be doubled. The acquisition of new buildings is subject to a 16% VAT or 4–7% for dwellings. Transfers of rural lands, sites, plots and existing buildings are exempted from VAT. Letting of commercial property is subject to 16% VAT.

18.6.8 Malaysia

Depreciation of land and buildings and capital allowances is generally not available for residential and commercial buildings. For buildings classified as 'industrial buildings' (for example, factory, warehouse, certain hotel buildings) and used for business or let out for rental, a capital allowance known as an 'industrial building allowance' can be claimed against business or rental income. The initial allowance is 10% and the annual allowance is 2% of the building cost. Apart from this, buildings constructed under an agreement with the Government on a build-lease-transfer basis qualify for allowances of 6%. The following types of buildings qualify for allowances of 10% per annum:

1. Buildings provided as living accommodation for employees by a person engaged in a manufacturing business, hotel/tourism business or approved service project.
2. Buildings used as a school or an educational institution approved by the Minister of Education or any relevant authority or for the purposes of industrial, technical or vocational training approved by the Minister.
3. Buildings used as a warehouse for storage of goods for export or for storage of imported goods to be processed and distributed or re-exported.

Capital allowances may be claimed on qualifying capital expenditure incurred on plant and equipment used in a business of letting property. The initial allowance is 20% and the annual allowance varies depending on the type of plant and equipment used. The current annual depreciation rates are: office equipment – 10%, furniture and fittings – 10%, general plant and machinery – 14%, heavy machinery – 20%, environmental protection equipment – 20%, and computer and information technology assets – 40%. Other tax-deductible expenses include property costs incurred, such as assessment rates, quit rent (termination payments), property insurance premiums and repairs and maintenance expenses. Costs which are capital in nature, such as stamp duty and legal costs incurred on the acquisition of property, are not deductible but would be regarded as forming part of the acquisition price of the property for real property gains tax purposes.

18.6.9 Singapore

Capital allowances are only available to those taxpayers who are deemed to be carrying on a trade, profession or business. The Inland Revenue Authority of Singapore generally regards non-commercial property letting as non-business and therefore capital allowances are not available in respect of residential property lets. Capital allowances may be claimed for qualifying capital expenditure on plant and machinery in commercial buildings. These claims can generally be made over the useful working life of the assets. In certain cases, accelerated claims can be made over either one or three years depending on the nature of the assets. Capital allowances can also be claimed in respect of capital expenditure on the construction of a qualifying industrial building. An initial allowance of 25% is made available together with an annual allowance of no more than 3%. The maximum number of years over which claims can be made is fifty years and subsequent owners of the property can claim the residue of the original expenditure on a straight-line basis within that fifty-year period. The residue of the original expenditure is restricted to the cost incurred by the last vendor and therefore any amount paid in excess of this by the acquirer would not qualify for capital allowances. A goods and services tax (GST), currently at a flat rate of 3%, is payable on the acquisition cost of a commercial property. If, however, an investor acquires commercial property with an existing rental income stream, this may be viewed as a 'transfer of a going concern' which is deemed an exempt supply and therefore not subject to GST.

GST is payable on rents derived from commercial property but is not a cost to the tenant if the business is registered for GST purposes. GST is not payable on the purchase of residential property and similarly is not levied on rents from residential property. Property tax is payable annually and is determined by the Property Tax Division of the Inland Revenue Authority of Singapore. Generally, it is based on the annual rental value of the property. The current rate of property tax is 12% for both commercial and residential properties, and is reduced to 4% for residential properties that are owned by individuals and are owner occupied. In calculating the net rental income, landlords are able to deduct all related outgoings and expenses incurred, including any interest payable on loans taken out to buy the property and any property taxes payable.

18.7 Feasibility/design considerations

In certain circumstances, particularly where varying depreciation allowances are applied for the various elements of a building, it can often be financially prudent for an owner to have the design team examine design options that may maximize their financial returns.

A typical example in Australia may relate to using carpet in lieu of floor tiles. Under current provisions, the tiles would be classified as part of the building capital works and depreciated at 2.5% per annum. However, carpet could be depreciated at up to 40% per annum. The intended period of ownership and the relative costs of the tiling and carpet would need to be considered in this situation. Another example relates to internal partitions in an office building. De-mountable partitions are depreciable as plant and articles at 20% per annum but fixed partitions are classified as part of the building works and depreciable at only 2.5%.

These factors are, however, only of concern to proprietors who intend to maintain ownership of the property upon completion rather than sell the building immediately.

18.8 Expert taxation advice

The following case studies on cost segregation provide an example of how the use of expert property taxation consultants may reap significant financial benefits for property owners. These types of benefits tend to be possible in countries where tax depreciation laws provide for individual assessment and interpretation of laws/rulings. In these countries, the taxation provisions are often complex and the use of expert consultants will enable owners to maximize their investment returns. Additionally, taxation laws and interpretations are typically changing all the time, making the use of expert property taxation advisers even more important.

In the US, entire commercial construction and acquisition costs are generally depreciated over a 39-year period, and residential properties over 27.5 years. However, a cost segregation study can identify costs that qualify for federal tax lives of five and seven years for personal property and 15 years for land improvement property, thus reducing a company's federal tax liability.

The term 'cost segregation' describes the process of splitting the acquisition cost or construction cost so that individual assets are allocated to the appropriate tax life. The tax benefits of moving project costs from 39 years to a shorter depreciable life are illustrated below (Grant, 2000):

	5 year property	7 year property	15 year property
Qualifying project costs	$100 000	$100 000	$100 000
First year depreciation	$20 000	$14 286	$6 667
First year depreciation under 39 years' life	$2 564	$2 564	$2 564
Increase in first year depreciation	$17 436	$11 740	$4 103
First year's tax savings (34% tax rate)	$5 928	$3 992	$1 395

A cost segregation study (CSS) carried out by an expert consultant can result in accelerated tax depreciation deductions, enabling individuals and companies that own real estate to lower their current income tax liability, thereby increasing current cash flow. Savings can be as high as 5% of the asset cost. On a $10 million property, for example, a 5% benefit would generate $500 000 in tax savings. These studies have generated millions of dollars in tax savings to owners of US real estate. However, given the complicated nature of the study, it requires a tax expert with an intimate knowledge of the tax codes, the relevant tax cases and a network of resources to maximize the benefits (Grant, 2000). Grant contends that the majority of property owners do not utilize such experts and, accordingly, do not reap the full taxation benefits that their ownership may confer.

Cost segregation studies can be performed on newly constructed facilities, existing facilities and renovation work to existing facilities. Savings derived from these studies flow directly to the bottom line in tax savings and cash flow.

The following case studies (Grant, 2000) further illustrate the tax savings benefits of cost segregation and the use of expert property taxation consultants.

18.8.1 Example 1

A company constructed an $11 million office building in 1988. During the first ten years of operations, depreciation expenses were originally calculated as $3 300 000. As a result of a cost segregation study performed in 1998, the company was able to increase its depreciation expense by over $1 600 000 during the next four years. This resulted in discounted present value tax savings and additional cash flow of more than $340 000 to the company.

18.8.2 Example 2

A $6 000 000 warehouse facility was put into service in 1997. As originally calculated, depreciation expenses during the first four years of operations were approximately $650 000. After a cost segregation study was performed in 1999, the company was able to increase its depreciation expense during the same four-year period by $225 000. This resulted in tax savings and additional cash flow of over $100 000.

18.8.3 Example 3

An $8 500 000 nursing home was constructed in 1987. As originally calculated, depreciation expenses during the first 11 years of operations were approximately $2 600 000. After a cost segregation study was performed in 1998, the company was able to deduct an additional $1 600 000 of depreciation spread over the next four years. This resulted in tax savings and additional cash flow of over $500 000 during the four-year period.

18.8.4 Example 4

An office building complex costing $48 000 000 was acquired in 1995. The owner made tenant improvements of $2 million to the facility over the ensuing two years. As originally calculated, the depreciation expense from 1998 through 2001 was $5 050 000. A cost segregation study that identified improvements such as millwork, wall coverings, kitchen plumbing, telecommunications wiring and supplemental air conditioning, increased depreciation expenses during that four-year period by $2 300 000. This led to tax savings and additional cash flow of over $700 000 to the owner.

18.9 Conclusion

The importance of tax depreciation for property owners and developers will depend on the particular taxation legislation applying in the relevant country. In countries such as the United States, Australia and the United Kingdom, taxation legislation makes tax depreciation an important consideration for property owners. However, in some other countries, the lack of depreciation provisions makes tax depreciation a non-issue. Nevertheless, given the increasingly global nature of construction activity and property acquisition, an awareness of tax depreciation is important for all property professionals.

References and bibliography

ATO (2001). Australian Taxation Office Website. <http://www.ato.gov.au>

Grant, D. (2000). 'How Cost Segregation Offers Substantial Tax Benefits to Real Estate Owners and Investors in Real Estate'. *Real Estate Issues*, 25(2), pp. 1-5, Summer.

IRS (2001). USA Internal Revenue Service Website. <http://www.irs.ustreas.gov>

Napier and Blakeley (2001). Napier and Blakeley Website. <http://www.napier.net.au>

Price Waterhouse Coopers (2001). Global Real Estate Tax Summary. <http://www.pwcglobal.com/extweb/pwcpublications>

Part 7

Value management

Value is often interpreted as cost-effectiveness. Under this definition, maximum value is achieved when the lowest cost is found. Cost should not be limited to initial expenditure (capital cost) but must extend to include recurrent expenditure (operating and finance cost) which is typically spread over a long time horizon. This concept is valid in cases where decisions are strongly investor-centred and where the performance of the project is set at an agreed level. This is not a common occurrence.

Not all decisions can be viewed in pure economic terms. Other matters, often of an intangible nature, are known to constitute a benefit but fall outside conventional financial analysis. It is necessary for decision-makers to have a balanced outlook and use judgement to ensure that the best course of action is taken. They are wanting value for money; in other words, they seek the solution that represents the best ratio of value (performance/quality/productivity) to cost (capital/operating/finance). The higher this rate the better, although the formal recommendation may not represent either the option with the highest benefit or the lowest cost.

Value actually means different things to different people, and is based on a complex set of personal ideals and desires that are not universally shared. Any value-based decision is therefore highly dependent upon who makes it. For this reason organizational decisions are best made by groups of people with different backgrounds and perspectives, ideally representative of specific areas of corporate responsibility.

Decisions related to assets usually involve lives of more than a few years and therefore attract a cash flow pattern that cannot be treated as a one-off payment. It follows that a long-term view must be taken in these cases so that all issues that arise from the decision are considered. This concept is called whole-of-life assessment. It is not restricted to tangible (monetary) issues but must also include the value of intangible benefits. Techniques have been developed to enable these complex evaluations to occur using an objective and rigorous process.

Part of the whole-of-life philosophy is having regard for environmental impact such as energy, waste, pollution, land use and material selection. Given that proper values are placed on these items, resultant decisions will reflect a true picture of proposed actions and enable balanced decisions to be made. This approach will provide projects that embrace sustainable development goals to be recognized. It is vital that facility managers understand the importance of sustainable development, not only in terms of the wider social responsibilities, but from the perspective of how it benefits the organization and its image in the community.

Value needs to be the focus of all strategic-related decisions regarding facilities. Value management is a process that has developed over many years and is now routinely practised in a range of industries as a tactic to improve quality, reduce cost and minimize risk. Therefore value management (VM) is a vital tool for the facility manager. It enables tangible and intangible costs and benefits to be collectively considered and options to be assessed against identified performance criteria. These criteria are weighted to reflect their comparative significance. Brainstorming, lateral thinking, functional analysis and whole-of-life assessment are all integral to the technique.

VM is actually a form of multi-criteria analysis (MCA). Put simply, this technique enables potential courses of action to be ranked on the basis of the criteria that describe the purpose or function to be achieved. In MCA there is no reason why each criterion cannot be expressed in different units. In some cases VM may help to measure performance for some criteria, but different approaches may be more suitable in other cases such as financial return, energy and extent of environmental impact. Techniques of this type have advantages over traditional cost-benefit analysis and discounted cash flow forecasts in that they incorporate intangible values as a formal part of the decision model.

Qualitative performance indicators such as productivity are clearly of interest at an organizational level, but the challenge is to show the link between facility decisions and productivity improvement in a way that confirms a causal relationship exists. Such studies must take into account tangible and intangible issues and measure value for money expectations. MCA is a suitable methodology to use in conjunction with VM tools like brainstorming. A multi-disciplined environment is appropriate to ensure that people can be involved in the decision process and bring a range of experiences to bear on the discussion. In many cases arrival at a successful solution is as much a function of organizational commitment as it is a good idea.

This part investigates the principles of value management as applied to facility performance. Chapter 19 looks at whole-of-life assessment and shows how sustainable development requires a long-term view inclusive of tangible and intangible values. Chapter 20 discusses the need to build value into decisions and how to identify solutions that represent value for money. Chapter 21 looks at the technique of multi-criteria analysis.

Value is an abstract concept, but forms a suitable context for all decision-making activities. Maximization of value is the goal, but should not be confused with maximization of financial return (or minimization of cost) which constitute only part of the total benefit. An appropriate balancing of performance criteria is necessary so that decisions reflect wisdom and embody socially responsible ideals and aspirations.

Whole-of-life assessment

19.1 Introduction

A 'whole-of-life' assessment methodology is necessary if the construction/property industry is ever to be serious about ensuring that development is sustainable. While there are assessment tools available for use by the industry, there is no connection between many of these tools, and often the results of different approaches can give contradictory advice. There is a clear need to identify and evaluate sustainable solutions to the facility problems at all stages of the property cycle.

The philosophy of sustainable development borrows freely from the science of environmental economics in several major respects. A basic component of environmental economics concerns the way in which economics and the environment interact. Recognition of the fact that the economy is not separate from the environment is fundamental to any understanding of sustainable development. The two are interdependent because the way humans manage economies impacts on environmental quality, and in turn the environment directly influences the future performance of economies.

The risks of treating economic management and environmental quality as if they are separate, non-interacting elements have now become apparent. The world could not have continued to use chlorofluorocarbons (CFCs) indiscriminately. That use was, and still is, adversely affecting the planet's natural ozone layer. Furthermore, damage to the ozone layer affects human health and economic productivity. Few would argue now that we can perpetually postpone taking action to contain the emission of greenhouse gases (GHGs). Our use of fossil fuels is driven by the goals of economic change and that process will affect global climate. In turn, global warming and sea-level rise will affect the performance of economies.

Definitions of sustainable development abound, since what constitutes development for one person may be neither development nor progress for another. Development is essentially a value word: it embodies personal aspirations, ideals and concepts of what constitutes a benefit for society. The most popular definition is the one given in the Brundtland Report: 'development that meets the needs of the present without compromising the ability of future generations to meet their own needs' (World Commission on Environment and Development, 1990).

This definition is about the present generation's stewardship of resources. This means that for an economic activity to be sustainable it must neither degrade nor deplete the natural resources, nor have serious impacts on the global environment inherited by future generations. For example, if greenhouse gases build up, ozone is depleted, soil is degraded, natural resources are exhausted and water and air are polluted, the present generation clearly has prejudiced the ability of future generations to support themselves.

Development implies change, and should by definition lead to an improvement in the quality of life of individuals. Development encompasses not only growth, but also general utility and well-being, and involves the transformation of natural resources into productive output. Sustainable development in practice represents a balance (or compromise) between economic progress and environmental conservation in much the same way as value for money on construction projects is a balance between maximum functionality and minimum life-cost. The economy and the environment necessarily interact, so it is not appropriate to focus on one and ignore the other.

Development is undeniably associated with construction and the built environment. Natural resources are consumed by the modification of land, the manufacture of materials and systems, the construction process, energy requirements and the waste products that result from operation, occupation and renewal. Building projects are a major contributor to both economic growth and environmental degradation and hence are intimately concerned with sustainable development concepts.

Having said that, there is probably no such thing as sustainable development for the general run of construction projects. But that is not to say that consideration of sustainability is a waste of effort. On the contrary, every project (new or existing) can be enhanced by consideration of whole-of-life methodologies. While most projects will consume more resources than they create, projects that are closer to sustainable ideals will increasingly deliver benefits to their owners and users and to society as a whole. Therefore if design can encompass assessment and decision-making processes that address sustainability goals, it is likely that over the long term the construction industry will be able to demonstrate a significant contribution to global resource efficiency.

19.2 The 'whole-of-life' philosophy

Whole-of-life assessment simply means taking account of all factors that affect a particular development over its effective life. Some of these factors can be assessed objectively, such as costs and energy usage, and others are more subjective, such as welfare enhancement and environmental impact. Most established methodologies focus on the monetary translation of 'benefits' (or advantages) and 'costs' (or disadvantages) over time, and often use discounting to bring these values into an equivalent present value for evaluation.

Although the concept of considering all the costs of a project may seem common sense, in current practice decisions are made largely, and unfortunately, on the basis of capital costs alone. This results in value for money being viewed in a narrow context and limits the possibility of achieving efficiencies in the employment of resources within society. While value assessment must additionally include factors such as functionality, aesthetics, environmental impact and financial return, operating costs remain a significant part of any asset decision. The design professions must be conscious of the importance of operating costs and seek out solutions in their practice that will not ultimately become financial burdens for their clients or others in the community. The ability to deliver value for money on projects whilst concentrating solely on capital cost issues remains a contemporary fallacy (Langston, 1999).

Past analyses of design solutions for building projects have concentrated on initial capital costs, often to the extent that the effects of subsequent operating costs are completely ignored. However, even in cases where a wider view of cost has been adopted, the discounting process has commonly disadvantaged future expenditure so heavily as to make performance after the short term irrelevant to the outcome, resulting in projects that display low capital and high operating costs being given favour. Thus design solutions that aim to avoid repetitive maintenance, reduce waste, save non-renewable energy resources or protect the environment through selection of better quality materials and systems, usually having a higher capital cost, are often rejected on the basis of the discounting process.

The monetary assessment of projects using a whole-of-life approach has been plagued with difficulties ever since the concept was first practised. Even today, after much research and development has led to most problems being laid to rest, there is still a persistent recounting of these difficulties in industry. The most commonly quoted difficulties comprise forecasting accuracy, lack of historical data, professional accountability, technology changes, capital versus operating budgets, and the underpinning theory of discounting (Ashworth, 1994).

It is true that forecasting the future is always likely to lead to variations in results between initial predictions and subsequent reality. But this does not add weight to the argument that a whole-of-life approach to monetary assessment is flawed. The solution to the problem lies in the routine use of risk analysis, whereby the uncertainty of future events can be evaluated against the robustness of assumptions to add confidence levels to final decisions. The bottom line is that uncertainty may be high, particularly where long time horizons are involved, but the risk of making a poor decision between various proposed courses of action may be quite low.

There is a lack of historical data about maintenance, energy usage, operating costs and life expectancies of materials and systems. Some useful resources do exist, but the ability to create a comprehensive database is problematic. The point is that even if a comprehensive database existed now, it would need to be used with great caution, since historical information is highly context dependent. For example, operating costs of buildings will differ due to the variable durability of materials and systems and the effects of their interaction with each other, the environment and the users of the building. Different occupancy profiles, hours of operation, location, environmental factors, management policy, quality of the original construction, premature damage and building age will all affect the integrity of the data being collected. Long time frames are involved in the establishment of data, and new methods and systems are always being added. Rather than focus on historical data, estimating operating costs from first principles is recommended, and context-specific databases will be formed from the routine comparison of predicted and actual performance.

Professional consultants, such as quantity surveyors and engineers, will not normally be accountable for the accuracy of their work in this area, since it may take many years to test whether their predictions are valid. But whole-of-life planning should not be considered as an estimate of actual cost, rather as a framework within which effective cost management processes can occur. Elemental targets and individual work items act as benchmarks for design and subsequent operation. Where cost over-runs occur, the cost management process can focus on the reasons why and on implementing strategies which will ensure that overall cost limits are not exceeded. Professional accountability, therefore, should be about issues of methodology, capability and process, not fortune-telling skills.

Changes in technology and social fashion occur regularly. It would be unrealistic to think that original design decisions will continue to represent good value for money until their life expectancy is reached. The cost management process must be able to adapt and re-evaluate new options as they appear in the marketplace, and this must be seen as another routine activity with any whole-of-life approach. Therefore systems must be used that support the concept of continuous cost management from concept through to demolition or disposal (cradle to grave). Technology changes are not problems, merely new pieces of information.

A classic difficulty is the artificial barriers that have been created, particularly in the public sector, between capital and operating budgets. Any whole-of-life approach relies fundamentally on the ability to spend more money initially in order to make overall savings, and if this cannot occur then the entire process becomes pointless. Therefore it is vital that clients remove these internal divisions and treat project budgets as inclusive of initial and recurrent cash flows. There is plenty of evidence of success in this area, even in the public sector, but more policy emphasis is required.

Discounting is generally regarded as a controversial process. This is not because there are unresolved theoretical issues that require further research, but rather that existing research has not adequately been disseminated to industry. One major misconception, for example, is that discounting leads to bias against future generations and therefore is not compatible with sustainable development goals. This assertion is incorrect provided the rate of discount is based on the weighted cost of capital to be used in the project. Discount rates in this context are merely a mechanism to incorporate financing costs (such as interest paid on borrowed funds or interest lost on use of equity) within the analysis. In the past, high discount rates have led to poor decisions (Langston, 1994).

There are other minor difficulties that sometimes appear in the literature, but the only real barrier these days to widespread implementation of a whole-of-life approach throughout the construction/property industry is the attitude of individual stakeholders.

Sustainability should be seen as an ideal goal or the ultimate in development efficiency. In almost all cases real sustainability is unachievable, but the degree of compliance indicates the success of the design in terms of society's expectations. The assessment of sustainable development must rely on a multi-criteria approach. It would be naive to assume that everything can be expressed in an equivalent monetary unit such as net present value, and yet there is a need for an objective and rigorous approach that can produce repeatable results.

The principles of sustainable development comprise environmental valuation, futurity and equity (Pearce et al., 1989). Environmental valuation means placing proper values on the consumption of renewable and non-renewable natural resources. Futurity means taking decisions in the context of the time period over which the development influences its surroundings. Equity is about fair treatment of groups within current society (intragenerational equity) and between present and future society (intergenerational equity). These three principles must underlie any whole-of-life assessment methodology.

19.3 Assessment tools

19.3.1 Environmental impact assessment

There is significant and increasing awareness of global environmental problems, in particular relating to global warming, ozone depletion, biodiversity, pollution and population growth. With this increasing awareness comes the realization that the potential impacts of proposed development activities need to be assessed and understood so that appropriate management and control strategies can be adopted.

Environmental impact assessment (EIA) has been developed and is used as a means to identify potential damaging effects of proposed developments and is widely accepted as an effective tool of environmental management. EIA was first introduced in the United States in 1969 under the National Environmental Policy Act (NEPA) for land-use planning. Environmental quality was identified as a national priority and the EIA procedure established by NEPA became a worldwide model for implementation.

Most countries now have EIA procedures in place and use them to assess projects that are environmentally sensitive (Gilpin, 1995). The definition of EIA may be described as 'a procedure for encouraging decision-makers to take account of the possible effects of development investments on environmental quality and natural resource productivity and a tool for collecting and assembling the data planners need to make development projects more sustainable and environmentally sound [and . . .] is usually applied in support of policies for a more rational and sustainable use of resources in achieving economic development' (Clark, 1989).

Environmental impact statements are undertaken usually by specialist consultants who may draw on other consultants for particular aspects of the evaluation. But it is the translation of the results into objective assessment and the decision-making processes that follow which are of primary interest to any whole-of-life methodology. Where environmental impacts are highlighted and form barriers to project implementation, solutions need to be found and the costs of these solutions established, otherwise a large amount of time and effort is expended for no result. The cost implications of environmental impacts often span long time horizons, and their assessment becomes closely linked with economic issues and government policy direction. For example, the impact of taxation on the financial viability of a project can be significant, and as governments exercise more control over development in an effort to reduce the consumption of scarce resources, so fiscal instruments will become even more influential.

Current research is focusing on the use of multi-criteria analysis (MCA) as an alternative to conventional economic evaluation (van Pelt, 1993). Problems with the application of shadow pricing and the monetary translation of externalities and intangibles have led to more interest in qualitative assessment. This technique is not dissimilar in concept to the form of value management adopted by the construction industry. A weighted matrix approach may be a more relevant tool to apply than discounted cash flow analysis if used effectively in a multi-disciplined team context.

19.3.2 Cost-benefit analysis

Cost-benefit analysis (CBA) is the practical embodiment of discounted cash flow analysis. There are two types of CBA: economic and social. Economic analysis involves real cash flows that affect the investor. Social analysis involves real and theoretical cash flows that affect the overall welfare of society.

Economic CBA is a technique for assessing the return on capital employed in an investment project over its economic life, with a view to prioritizing alternative courses of action that exceed established profitability thresholds. It uses discounted cash flow analysis to make judgements about the timing of cash inflows and outflows and envisaged rates of return. Most experts agree that discounting is fundamental to the correct evaluation of projects involving differential timing in the payment and receipt of cash. Accounting rate of return and simple payback methods, which do not consider changing time value, are quite inadequate substitutes and may produce misleading advice.

The two most common capital budgeting tools used as selection criterion in CBA are net present value (NPV) and internal rate of return (IRR). Both rely on the existence of costs and benefits over a number of years, and lead to the identification and ranking of projects that are financially acceptable for possible selection. NPV is the sum of the discounted values of all future cash inflows and outflows. If the NPV is positive at the end of the economic life, then the investment will produce a profit with reference to the discount rate selected. If the NPV is negative, then the investment will produce a loss. Mutually exclusive projects should be selected on the basis of the magnitude of the NPV, provided it is positive. IRR is defined as the discount rate that leads to an NPV of zero. Depending on the cash flow pattern, some projects may exhibit multiple IRR while others may have no rate at all.

Social CBA uses the concept of collective utility (aggregate efficiency) to measure the effect of an investment project on the community. Externalities are assessed in monetary terms and included in the cash flow forecasts, even though in many cases there is no market valuation available as a source for the estimates. Nevertheless the technique is useful in that it attempts to take both financial and welfare issues into account so that projects delivering the maximum benefit can be identified.

Because social CBA is directed at maximization of collective utility, it is generally employed as an investment technique by government agencies. Furthermore, the amount of research and estimation that the technique demands means that social analysis is feasible only on large and complex projects, typically those involving infrastructure such as transport, healthcare, water irrigation, energy supply and the like.

More frequently the technique is being called upon to evaluate environmentally-sensitive projects (Field, 1997). It has, however, come under attack from many conservationists because it can justify investments that cause damage to the environment provided that this damage is outweighed by an increase in capital wealth. The quantification of externalities clearly has a great impact on the outcome. In addition, distributional effects are often overlooked, as the acceptance criterion is related to overall utility improvement and hence does not recognize whether this benefit is evenly shared. For this reason projects that are likely to involve an uneven distribution of benefits should be the subject of a separate study designed to expose such effects.

Social CBA is the primary tool used in environmental economics, yet the technique still has significant problems when applied to the evaluation of environmentally-sensitive projects. The most significant problem concerns the fundamental concept of aggregate efficiency. In its basic form, social CBA does not adequately account for environmental issues. A major reason for this is the Kaldor-Hicks principle that underlies CBA theory. It states that any type of cost to society is acceptable as long as a project generates greater benefits. Therefore environmental damage is acceptable if benefits, such as increases in capital wealth, are valued more highly.

No CBA principle prescribes that part of the benefits should actually be reinvested in measures to avoid or compensate for environmental damage, and therefore substitution between capital and environmental wealth is perfect. Yet from a sustainability point of view, it would be preferable to limit environmental damage in absolute terms once a particular threshold has been reached regardless of the increases in capital wealth that may be anticipated.

Externalities often pose significant problems in their monetary translation, and shadow pricing may become necessary. These costs and benefits can be significant, and unrealistic assessment can easily distort results. Where externalities are beyond assessment altogether, and hence become intangibles, important aspects of the project may be neglected. Environmental costs and benefits may often be considered intangible, and clearly this fails to adequately address the sustainability question within the boundaries of the technique.

19.3.3 Life-cost studies

The measurement of development costs is commonly undertaken on a capital cost basis. Projects are analysed in respect of their likely construction costs, to which might be added land purchase, professional fees, furnishings and cost escalation to the end of the construction period. Budgets similarly address initial costs, and the planning and control processes they foreshadow normally do not extend beyond hand-over. Developers are often not interested in long-term costs that will be borne by one or more future owners, and as yet there is little government regulation that makes either presentation or consideration of long-term costs commonplace (Langston, 1997).

The need to look further into a project's life than merely its design or construction is obvious, since operating costs usually outweigh expenses involved in acquisition. A whole-of-life approach takes into account both capital and operating costs so that more effective decisions can be made and thus satisfies the primary objective of clients and agencies, that of achieving value for money. It delivers the following tangible benefits:

1. Assesses alternative courses of action.
2. Investigates the sustainable deployment of limited natural resources.
3. Identifies cost-effective designs.
4. Provides a balance between capital and operating budgets.
5. Calculates operating cost cash flows over the life of the project.
6. Enables the financing of future commitments.
7. Enables continuance of cost management throughout the stages of design, construction and occupation.
8. Facilitates meaningful data collection and feedback from existing buildings.
9. Provides information for financial planning and analysis of future expenditure.
10. Evaluates the burden being placed on future generations to maintain buildings.
11. Monitors the success of design choices against estimated targets and highlights possible areas of improvement. (Langston, 1991a)

These advantages build on those already implicit in the rigour of the cost management process. Nevertheless, the challenge ahead is to ensure that short-term financial considerations do not prevent the pursuit of long-term benefits.

The ability to balance capital and operating costs is crucial to the effective use of any whole-of-life methodology. The concept is simple and for most materials and systems holds true: spending more money initially will lead to lower operating costs over the life of the project. Quality assurance assumes greater importance as poor workmanship may lead to premature failure of expensive and otherwise long-lasting components. While this concept is often applied to issues of maintenance, repair and replacement, a similar relationship can be expected with energy. However, in some cases substantial savings can be delivered by using low technology solutions that are inexpensive to construct and operate. The greatest value for money savings may well lie in passive heating and cooling systems and the increased use of natural light and ventilation.

During the pre-design stage it is important to establish a realistic framework for financial management and to investigate options that have the promise of delivering maximum value for money. A value management process is often employed to bring together aspects of function and cost so that an effective balance can be struck. This requires the application of detailed knowledge derived from past experience and investigation in a form that can assist new conceptual designs. Investment analysis is the economic decision-making activity during the pre-design stage. Life-cost planning applies to the project's design development and life-cost analysis applies to both construction and subsequent occupancy. These three linked activities rely on a whole-of-life approach to project evaluation that takes proper account of operating costs over time.

Life-cost studies comprise both measurement and comparison activities. Measurement relates to the total cost of a particular design solution and involves the establishment of expected performance targets and their monitoring over time. Comparison relates to a number of alternative design solutions and in addition to total cost (and in many cases differential revenue) must consider the relative timing of receipts and payments. Most of the literature concentrates on the latter activity and hence gives the impression that discounting plays an absolute role in life-cost studies. But this conclusion is disputed, and it is recommended that measurement activities are important and their presentation in terms of real value is more meaningful than the use of discounted value (Langston, 1991b).

Research into life-cost studies may have uncovered and resolved many of the conceptual problems and misunderstandings that have haunted the technique for some time, but it has also highlighted the enormous application which exists in the assessment of sustainable development. Life-cost plans (expressed in present-day dollars) can be used to support essential facility management processes over the life of the project. Budgeting, cash flow forecasting, data collection and design feedback are direct and tangible outcomes from the quantification of life-costs. These are measurement activities and do not involve discounting. The life-cost plan becomes a framework that supports knowledge acquisition and encourages the adoption of a total cost attitude in project design.

Although the capital cost fallacy remains, there is an increasing interest in life-cost considerations. When the technique became popular in the 1960s in the UK it was because there was general concern about the expense attached to the maintnance of an ageing building stock and the impact this would have on the growth and prosperity of the country in the years ahead. The technique emerged again in the mid-1970s when the oil crisis created doubts over the ability of building owners to operate energy-intensive buildings into the future. Today the technique is back on the agenda under an environmental banner. Sustainable development is a global issue and one which may turn out to be a more significant driving force for enabling life-cost considerations to become a routine part of professional practice.

Life-costs are of a significant nature and deserve consideration commensurate with their effect. Their absence from financial decision-making in the past has largely stemmed from the difficulties of applying these techniques and the tendency by professional advisers and building owners to ignore the impact of long-term expenditure.

19.3.4 Total energy analysis

Buildings 'consume' a significant proportion of the world's annual energy production in their construction, operation, maintenance, renovation, demolition and disposal. As a consequence of this energy consumption, the greenhouse gas emissions associated with fossil fuel use have a significant environmental impact. From both an environmental and economic standpoint, it is logical that sustainable development principles seek to minimize total life cycle energy inputs and thereby make more efficient use of scarce energy resources.

Strategies developed in the past twenty years to reduce the operational energy consumption of buildings have largely focused on the potential energy savings associated with increasing the thermal efficiency of a building's envelope. Whilst this is a positive advance, it addresses only one part of the total energy balance of a building's life cycle. The passive solar design principles that have evolved in response rely on the use of 'heavy' building materials, such as brick, concrete and stonework. It is now known that these materials also have a heavy environmental burden due to their high embodied energy content. Thus the operational energy savings achieved by implementing passive solar techniques have in reality largely been offset by increases in embodied energy in the construction phase of a building's life cycle (Mackley, 1998).

Embodied energy is defined as the quantity of energy required by all of the activities associated with a production process, including the relative proportions consumed in all activities upstream to the acquisition of natural resources and the share of energy used in making equipment and supporting functions (Tucker and Treloar, 1994). In order to estimate the embodied energy of a building, the elements must be disaggregated into their principal materials. The mass of each of the principal materials is calculated and is then multiplied by its embodied energy intensity to give a total figure which is generally expressed in gigajoules per tonne (Gj/t).

Energy analysis is the method used to measure the energy involved in a particular process. The establishment of what is known as a system boundary is essential to the ability to effect the identification and quantification of energy flows. Once clearly established, the system boundary must be strictly observed to ensure valid measurements. The application of life cycle principles to quantify energy flows through buildings is a relatively new but rapidly growing application. It involves the simultaneous comparison of a group of alternative projects to enable the selection of the most favourable alternative, such as the one with the most efficient net energy balance over its life, under a given set of defined conditions. The application of energy life cycle analysis is seen as an holistic and sustainable approach to the development process, which ensures that benefits (both ecological and economic) flow to both current and future generations through pursuit of energy conservation principles.

The sustainability concept implicitly suggests that new development should be designed so as to minimize its impact on the consumption of resources required for construction and future operation. While some control may be able to be exercised over this matter within normal investment appraisal activities, it is argued that additional control is necessary. The process of energy auditing can provide this control, but to be successful it must become a mandatory part of development approval.

Energy auditing can be divided into two parts: estimation and verification. Estimation involves the establishment of energy targets based on the design of the project and projections of the level of occupational usage. These would initially concentrate on operating performance but may in the longer term be extended to include the energy implied in resource extraction, processing and manufacturing. Verification involves the monitoring of actual energy use and comparison against the original targets. Action may then be necessary to correct situations where energy usage is shown to be excessive.

Energy auditing can be integrated into the development approval process with relative ease, since planning authorities already have the jurisdiction over whether new development is acceptable. Each development application should be accompanied by a statement of annual energy expectations appropriately classified into type and cause. But verification is a different kind of problem. It requires an independent audit of actual performance in much the same manner as an accountant might verify the accuracy of financial statements and taxation liabilities. Therefore an annual submission to a central authority setting out total energy consumption is necessary. The building owner would have the opportunity to balance the total energy usage by making savings in some areas to compensate for over-runs in others.

19.4 Proposed methodology

19.4.1 Phase one: development management

This is the most critical phase of the building cycle since the decisions that are made here will constrain the choices that follow. The development management phase is defined as covering all pre-design processes, and specifically includes the evaluation of competing investment opportunities, the investigation of environmental impact and the comparative assessment of sustainability.

The evaluation of competing investment opportunities is undertaken using conventional cost-benefit analysis. The cash flow should be relevant to the owner or providing authority and therefore for private sector clients may be limited to tangible costs and benefits over the expected holding period, while for public sector clients may be extended to include externalities over the expected economic life. Each alternative is judged on the basis of net present value (NPV), and any alternatives that show a negative NPV are instantly rejected. The risk attached to significant variables such as discount rate and time horizon must be explored. The analysis is recalculated for public sector clients to also show the NPV with externalities excluded, so that an indication of client commitment can be ascertained. Each NPV is converted into a benefit-cost ratio (BCR) for later use.

Projects that have likelihood of significant environmental impact require a separate environmental impact statement (EIS). This activity can occur simultaneously with the cost-benefit analysis, and may in fact become one document. The EIS identifies the issues that require consideration by planning authorities and which must be satisfactorily addressed before planning permission is granted. Most projects have a negative impact on their environment, but what is critical is that where these impacts are significant then measures are needed to reduce them to acceptable levels or to provide tangible compensation. This is the responsibility of planning authorities, and clients will only be allowed to proceed with their projects where these issues have been properly addressed.

All development applications would also need to substantiate that their energy targets are within the limits set down by the regulatory authority in order for approval to be granted. This can be achieved using a life-cost approach based on element and sub-element classifications, but rather than exploring the targets in terms of cost they would be expressed in units of energy (such as Gj/m^2 or a suitable alternative). Subsequent failure to meet established energy targets would attract a financial penalty for the building owner that could be collected as part of the taxation system. Existing buildings would also need to comply for reasons of equity and fairness, but perhaps would be given dispensation through higher energy limits.

The comparative assessment of sustainability indicates which of the acceptable alternatives should be selected. It involves the use of multi-criteria analysis to combine key attributes (expressed in units that best represent their nature) together into a single decision criterion. The suggested approach is to calculate an index of performance as described by Ding (1998):

$$\text{sustainability index (SI)} = \frac{\text{benefit-cost ratio} \times \text{social benefit}}{\text{total energy} \times \text{environmental damage}}$$

where:

1. Benefit-cost ratio (ratio: 1) is calculated as the discounted project income ($) divided by the discounted life-cost ($) when measured over the economic life of the project. Social and environmental factors are not monetarized, so the ratio is reflective of normal economic CBA. The higher the ratio the better the result. A ratio of 1 indicates the financial break-even point.
2. Social benefit (value score) is calculated using a VM-style weighted matrix approach. Social benefits (externalities and intangibles) are identified and weighted, and the ability of each alternative to meet these criteria is scored. The higher the score the better the performance of the project.
3. Total energy (Gj/m^2) is the sum of the embodied energy used in construction and maintenance, plus the operational energy 'consumed' over the project's economic life. The lower the total energy content the better the result.
4. Environmental damage (% risk) is the likelihood of environmental disturbance being incurred over the economic life of the project. It can be assessed using a risk analysis strategy or derived from an environmental impact statement. The lower the probability of environmental damage the better.

The index enables relative ranking of project options. Using this innovative approach, projects can be selected on the basis of their level of sustainability in comparison with other options. The highest index indicates the best balance (value for money) and would therefore govern the final choice of project. While this may not be the project that generates the greatest financial return, it will be the most sustainable project from the list that returns a positive NPV. The characteristics of this project are then carried forward to the next phase.

19.4.2 Phase two: project management

The project management phase is defined as covering all design and construction processes, and specifically includes the optimization of design choices, measurement of expected performance and the efficient execution of work on site. This is the phase within which most design consultants are normally engaged.

The optimization of design choices is achieved by undertaking value management studies on significant areas of the project during its preliminary design. This is a mini form of the sustainability index concept mentioned earlier. Options are scored against a list of weighted performance criteria, and the multiplication of score and weighting give rise to a value score for each alternative. The performance criteria should be exclusive of any issue that can be objectively expressed in life-cost terms (for example, initial cost, durability, ease of maintenance, etc., should not be criteria). Each alternative should also have its life-cost calculated. The comparisons are expressed in discounted (present) value terms. Taxation considerations, where relevant, and residual values are included in the comparison. Risk analysis is used to determine the impact of major variables such as discount rate and time horizon. The final judgement is made by dividing value score by life-cost. This creates a value for money index and is a statement of balance between maximum utility (benefit) and minimum resource input (cost), where the higher the number the better value for money. This is a useful method for making informed decisions about complex problems.

The measurement of expected performance is carried out by producing a life-cost plan expressed in real (undiscounted) value. The cost plan is one of the principal documents prepared during the project management phase. Costs, quantities and specification details are itemized by element (or sub-element) and collectively summarized. Measures of efficiency are calculated and used to assess the success of the developing design. The elemental approach aids the interpretation of performance by comparison of individual project attributes with similar attributes in different projects, and forms a useful classification system. Life-cost plans differ from traditional capital cost plans only in the type of costs that are taken into account and how these can be expressed and interpreted. This information can be used to draw conclusions about cost distribution, cash flow and overall financial commitment, and forms a set of targets against which future performance can be compared.

The construction process itself should not be ignored, as issues such as waste minimization and pollution can impact on overall sustainability achievements. Systems must be established to collect and sort waste for the purpose of recycling, to minimize air and water pollution to surrounding areas and to ensure that compliance occurs. While financial return may be an incentive for the contractor, it is ultimately the role of environmental protection authorities to legislate and control relevant onsite activities.

19.4.3 Phase three: facility management

The facility management phase is defined as covering all operational processes, and specifically includes the review of actual performance in the light of established targets, enhancing worker productivity, and the continual reaction to new information and external inputs.

The review of actual performance in the light of targets established in the life-cost plan makes possible a degree of control previously unknown. It includes distributing actual costs amongst the appropriate elements (or sub-elements) upon which life-cost plans are normally based. Composite rates can also be averaged from real projects to provide a valuable guide to the estimation of new projects. Analysis of life-costs can form an essential element of overall cost management by highlighting the ways in which potential cost savings in existing buildings might be achieved. For example, it might be better value to prematurely replace an expensive building component with a more efficient alternative than to simply continue with the original decision until its useful life has expired. Prudent control requires that actual and expected performance be constantly compared.

If a life-cost approach is to be effective in reducing the running costs of existing buildings it is necessary that these running costs be continually monitored. Monitoring involves the recording of actual performance of a particular project in a form that facilitates subsequent life-cost planning and management activities. Such performance can be compared against the cost targets and frequency expectations given in the life-cost plan. Areas of cost over-run or poor durability can be explored, potential improvements identified, and better solutions implemented. Data from different projects reflects the specific nature of those projects, their locations and occupancy profiles. When measured overall, no two buildings will have identical running costs, nor will the running costs for any specific building be the same from one year to another. Data collection is still a useful exercise, particularly for monitoring activities, but obtained results should be applied to other situations with caution.

Worker productivity is another aspect of sustainable development that applies to the facility management phase. It involves maximizing the productive output of employees through good design and internal climate. The impact of such issues on the financial bottom line of an organization can be staggering. Occupancy costs (or functional-use costs) can far outweigh capital costs, energy, maintenance and other operating costs combined. Environmentally-friendly practices such as recycling and reduced consumption of paper can also be of importance.

The expected performance targets must be continually revised by new information (such as better products, technology advances, etc.) and external inputs (such as strategic priorities, competition and legislation). This means that assessment processes normally in previous phases can all be applied in a specific context to the management of existing facilities. Therefore cost-benefit analysis, environmental impact assessment, life-cost studies and total energy analysis all have application here. In particular, energy auditing is a valuable tool that can be used to seek further improvements in operating performance.

19.5 Conclusion

A collection of traditional and innovative approaches can be formed into a useful methodology for the objective assessment of sustainable development. A whole-of-life attitude is essential to success in this endeavour. The property cycle is described as comprising development management, project management and facility management phases that impact on the assessment of sustainability. By connecting the individual tools and establishing a framework within which they operate, it is hoped that the construction/property industry will adopt them in a more practical and routine manner, and hence make a tangible contribution to resource efficiency and increased quality of life worldwide.

References and bibliography

Alexander, K. (1996). *Facilities Management: Theory and Practice*. E. & F.N. Spon.

Ashworth, A. (1994). *Cost Studies of Buildings* (2nd edition). Longman.

Atkin, B. and Brooks, A. (2000). *Total Facilities Management*. Blackwell Science.

Barrett, P. (1995). *Facilities Management: Towards Best Practice*. Blackwell Science.

Clark, B. D. (1989). 'Environmental Assessment and Environmental Management'. In Proceedings of the 10th International Seminar on Environmental Impact Assessment and Management, University of Aberdeen.

Cotts, D. G. (1999). *The Facility Management Handbook* (2nd edition). AMACOM.

Ding, G. K. C. (1998). 'The Influence of MCDM in the Assessment of Sustainability in Construction'. In proceedings of 14th International Conference on Multiple Criteria Decision Making, Charlottesville, June.

Field, B. C. (1997). *Environmental Economics: An Introduction* (2nd edition). McGraw-Hill.

Gilpin, A. (1995). *Environmental Impact Assessment (EIA): Cutting Edge for the Twenty-first Century*. Cambridge University Press.

Langston, C. A. (1991a). *Life-Cost Procedures Manual*. Department of Public Works, Sydney.

Langston, C. A. (1991b). *The Measurement of Life-Costs*. Department of Public Works, Sydney.

Langston, C. A. (1994). 'The Determination of Equivalent Value in Life-Cost Studies: An intergenerational approach', PhD Dissertation, University of Technology, Sydney.

Langston, C. A. (1997). 'Life-Cost Studies'. In *Environment Design Guide*, Royal Australian Institute of Architects.

Langston, C. A. (1999). 'The Capital Cost Fallacy'. In *Building in Value: Pre-Design Issues* (R. Best and G. de Valence, eds). Arnold.

Mackley, C. J. (1998). 'Life Cycle Energy Analysis of Residential Construction: A Case Study', M.Build(CE) Dissertation, University of Technology, Sydney.

McGregor, W. and Then, D. (1999). *Facilities Management and the Business of Space*. Arnold.

Park, A. (1994). *Facilities Management: An Explanation*. Macmillan.

Pearce, D. W., Markandya, A. and Barbier, E. B. (1989). *Blueprint for a Green Economy*. Earthscan Publications.

Rondeau, E. P., Brown, R. K. and Lapides, P. D. (1995). *Facility Management*. John Wiley & Sons.

Schulze, P. C. (1999). *Measures of Environmental Performance and Ecosystem Condition*. National Academy Press, Washington.

Stevens, D. (1997). *Strategic Thinking: Success Secrets of Big Business Projects*. McGraw-Hill.

Tompkins, J. A. (1996). *Facilities Planning* (2nd edition). John Wiley & Sons.

Tucker, S. N. and Treloar, G. J. (1994). 'Variability in Embodied Energy Analysis of Construction'. In proceedings of CIB TG 16 Sustainable Construction Conference, University of Florida, November, pp. 183-191.

van Pelt, M. J. F. (1993). *Ecological Sustainability and Project Appraisal*. Avebury.

World Commission on Environment and Development (1990). *Our Common Future* (Australian Edition). Oxford University Press.

Building in value

Rick Best and Gerard de Valence

20.1 Introduction

In all manner of daily activities individuals and corporate entities try to get value for their money, usually within some fairly clear budgetary framework. This is true of a person looking for somewhere to buy lunch; it is equally true of an organization planning on spending money on the renovation or maintenance and use of a building. In both cases those who authorize the expenditure will be looking at how much is spent, what return they can expect in exchange for their money, and how well satisfied they will be with the outcome.

The person looking for lunch will weigh up a number of considerations including how much they are willing to pay, the physical appearance of the restaurant, how hungry they are, etc., while the organization's facility manager will have to juggle a much greater number of considerations such as corporate image, budgetary constraints (generally imposed by others), occupant comfort, worker satisfaction and productivity, work practices, modes of work (individual and/or group), corporate hierarchy, the indoor environment – the list is almost endless. In any event, those involved will be hoping to achieve maximum 'value for money'.

20.2 Value for money

The three basic parameters which influence the success or otherwise of a building project are time, cost and quality. Generally speaking, the success of any project which we initiate will be gauged in terms of these three factors – in simple terms we hope to achieve an outcome that is completed in the shortest possible time, for the lowest possible cost and which is of the highest quality. Changes in any one of these factors will influence the other two – higher quality will generally mean higher cost and may also require more time to complete (although this is not necessarily true, for example some components of very high quality may be easier to install or more compact and therefore take less time), or the time taken to complete may be compressed but only by the use of additional labour or plant, or perhaps the introduction of longer working hours with consequent financial penalties for overtime, hiring charges and so on.

The value which is achieved depends on balancing these three factors so that cost is kept as low as possible while time and quality are not compromised unduly. Value, however, is a very subjective consideration and certainly has different meanings for different people, for example to a collector a piece of modern sculpture may have great value, while to many people the same piece of sculpture may appear to be a pile of 'valueless' junk albeit one which may fetch a healthy price on the open market. So where is the value? Is it an intrinsic quality of the artefact or simply an expression of what someone will pay to acquire it?

Often the value of an item is related to its relative availability and demand: when abundant supplies of a commodity are available or there is little demand for it, then it has little value. In contrast, when a commodity is in short supply and/or demand is high the value increases. Value may also be expressed as a ratio of functionality to cost – greater functionality achieved for less cost thus means better value. Alternatively, value may be considered in terms of quality in relation to cost, and so maximum value may be obtained when a required level of quality is achieved at least cost, or by achieving the highest possible level of quality for a given cost, or from a balance of the two.

Value, from a corporate viewpoint, has different meanings in different situations. There are basic differences which depend on the nature of the company's interest in a facility – the most obvious is whether the facility is occupied by the company that owns it, or it is let to tenants.

Owner-occupied buildings function as production inputs into the business activities which their owners are involved in, whether it be manufacturing, processing, service provision, trading in goods, knowledge or information, or some other pursuit. In any of these cases owners gain value from their buildings by virtue of the business functions which they accommodate, support and facilitate. Facility managers, in this case, are charged with the responsibility of maintaining buildings so that they provide the maximum possible level of amenity that will enable their company's business functions to be carried out at the highest possible level.

Leased buildings, on the other hand, are simple revenue earners for their owners. Value in this instance is more straightforward, generally a function of income balanced against expenditure. For the facility manager, maximizing value revolves around repair and renovation of buildings so that they continue to generate a satisfactory level of revenue. This requires buildings to be maintained in a condition that keeps them suitable for existing tenants and attractive to prospective tenants, with a view to maintaining viable occupancy levels and a reliable income stream. Typically, it is the tenant's responsibility to create and maintain the level of amenity in their leased workspace through the decisions on the quality of the fit-out (furnishings and finishes) and the style of layout (workspace per employee and management structure).

20.3 Value in buildings

A building may be more valuable than another because it possesses any or all of a number of attributes or because it better satisfies the needs of its users. These include:

1. *Fitness for purpose*. This may be a function of the original building design or the result of well-planned adaptation or refurbishment of an existing facility. Regardless of how it is achieved, a building that serves its intended purpose well, accommodating all the activities and functions required by its owners or users in a practical and economic manner, is a valuable asset. It is self-evident that any building will be expected to perform the basic functions such as exclusion of external climatic factors (wind and rain), protection from pests and intruders and security for the people and goods which it houses.
2. *Flexibility*. Organizations change over time and the scope and nature of their activities change as well. Buildings which are flexible in terms of spatial layout, provision of engineering services, and suitability for adaptive reuse can continue to be of value to their owners or users, whereas buildings which are rigidly designed to accommodate a very specific activity or organizational structure may be of little value when the nature of the organization's business or structure change.
3. *Durability*. Solidly constructed buildings may continue to be valuable assets for centuries, however there are relatively few commercial facilities which are expected to continue in use across the same time spans that other sorts of buildings, including public buildings such as churches, institutional buildings such as university colleges, and houses, may be expected to remain in use. In fact, modern commercial premises may have economic lives of as little as twenty-five years, and undergo several large-scale refurbishments within that time. Buildings which may be expected to require extensive remodelling or refitting at relatively short intervals, in order to remain economically viable, need not be of particularly durable construction given that refurbishment will be driven by market forces that require buildings to show a 'new face' at regular intervals so that they remain attractive to prospective tenants.
4. *Uniqueness*. A building may be valuable because of some attributes that it possesses which sets it apart from its competitors. Such attributes may be related to site or position – a commanding view or location adjacent to a major commuter interchange that provides enhanced business opportunities for tenants can increase building value substantially. Historical or symbolic connections may also be important, or a unique appearance which makes the building an easily identified symbol of the organization that occupies it.
5. *Structural integrity and standard of finish*. A well-made building which exhibits high standards of materials and workmanship will be a valuable asset, and may be expected to retain its value as long as it has utility – regardless of how well made it is, a building which cannot fulfil the functions required of it or which is prohibitively expensive to operate and maintain may have little value.
6. *Acceptable indoor environment*. Modern buildings in particular are expected to provide comfortable indoor environments (in terms of ambient noise levels, temperature, humidity, ventilation and air flow) as well as promoting the health and well-being of their occupants. In addition, organizations are now looking with increased interest at the effects which indoor environments have on the productivity of their workers. Independent control of workspaces through such measures as individual ventilation and lighting control, separation of personal and communal space, provision of workspaces that are tailored to differing modes of

work, for example the 'den', the 'hive', the 'cell' and the 'club' identified by Duffy (1998), acoustic treatments, glare factors, proximity to windows, opportunities for the application of natural ventilation and daylighting strategies – these are just some of the possibilities which now exist for potential enhancement of worker productivity.

7. *Low cost*. This includes purchase or construction cost, maintenance costs, refurbishment costs, operating costs (for example, energy, cleaning) and occupancy costs. Over the economic life of a commercial building the most significant by far are the costs directly associated with employees: salaries, sick leave, recreation leave and the like (Romm and Browning, 1994). Other costs are better assessed using life-cost methods to identify and select alternative strategies that will minimize costs over time (Langston, 1999).
8. *Environmental impact.* Buildings which have less deleterious effects on the natural environment, through material selection, reduced energy demand and design for reuse (of materials and components, as well as a result of design which allows easy adaptive reuse of whole buildings) are becoming increasingly attractive in the marketplace. Partly this is driven by a new sense of environmental responsibility which is becoming a significant force in many facets of modern life, but it is also driven by a perception that there are real economic benefits available at a number of levels. Many organizations are now using a 'green' image as a promotional tool, and appearing to be environmentally friendly is now a common advertising strategy. It is a happy coincidence that many strategies which can reduce the environmental impact of buildings also produce economic benefits, for example, the adoption of daylighting and natural ventilation in offices has been shown to increase employee satisfaction and reduce the incidence of building-related illness, which in turn increases both the quantity and quality of the output of the occupants (Best, 1999; Smith, 1999).
9. *Revenue generation*. This may be in the form of rent received or it may be measured in terms of productive output – in either case building owners are expecting to make a profit from activities associated with their buildings.

Thus the value of a building may be enhanced by raising the level of satisfaction of each of the attributes listed above.

Increasingly, clients are assessing the value of their building in different ways. While a developer may measure the value of a new building in simple profit-making terms, those clients who build with a view to maintaining an interest in the completed project, either as rental property or to occupy themselves, are looking at the concept of the value embodied in these assets in new ways. There are a number of key areas to be considered: life-costing, environmental performance, indoor environmental quality and building services are amongst the most important. These factors are highly interdependent – for example, longer lasting materials or components may cost more initially but save money over time.

The selection of materials and components can be based on a whole-of-life analysis of available alternatives with value over time being the basis for decision-making. It is not uncommon, however, that such materials and components have lower environmental impacts (being more durable) and both cost less to run (through improved energy efficiency) and improve the quality of the environment inside the building (by providing a more comfortable and controllable environment) thus leading to increased productivity in the workplace. It is only fairly recently that compelling evidence has begun to emerge which demonstrates the degree to which these factors can affect the success or failure of building design, from a functional point of view, rather than in terms of aesthetics.

Clients are beginning to ask questions about the environmental impacts of different materials and techniques, and the possibility of reducing operating costs through more energy-conscious design. As a result there is a growing perception that better value for money can be achieved in the longer term if these concerns are addressed in the earliest stages of the procurement process.

20.4 Facility managers and building value

While facility managers are, by definition, generally involved in managing buildings that are complete and in use, they can nevertheless provide much useful input at all stages of the procurement process and so influence the value that is realized by building owners.

Facility managers are in a position to have profound effects on achieving and maintaining value in the buildings for which they are responsible. Their influence may be felt from the earliest stages of building procurement, whether a building is being bought or built, and onwards throughout the building's life. It is unfortunate that of the many people who are consulted during the design and construction of a building, the facility manager, who will ultimately have a much longer association with the building than any other individual, is seldom consulted. In the same way that mechanical engineers are all too often presented with a design into which they have had no input whatsoever and expected to 'make it work' (Lovins and Browning, 1992), so facility managers inherit building solutions to which they, too, have had no opportunity to contribute. Like the engineers, they must 'make it work' and accept that any number of parts and components of the building are not what they would have chosen based on their prior experience. This often presents facility managers with major difficulties of maintenance, management and replacement. A simple example is the selection of light fittings located in inaccessible parts of a building, in an atrium ceiling for instance, which incur substantial cost when globes need to be replaced or fittings require cleaning. These fittings may be selected for appearance or on the basis of lower initial cost, but can be much more expensive in the longer term because of accessibility problems.

Strategic facility management, however, goes well beyond questions of light globes and ladders. Through their intimate association with the buildings that they maintain, and with their occupants, facility managers are in a unique position to gather a broad knowledge of the factors which affect occupant comfort and satisfaction, durability of building fabric and components, the interaction between building occupants and their workplace and a host of other factors which determine a building's overall success or failure in terms of the satisfaction of corporate goals. Post-occupancy evaluation, which is essentially a survey carried out to gather opinions about a building's success or otherwise from those who occupy and use it, is not often carried out in spite of the apparent benefits which could be gained from a more thorough understanding of why some buildings function more successfully than others. A perceptive facility manager, however, even without undertaking any formal building evaluation, can develop a very useful body of knowledge about what works and what doesn't and apply this knowledge to building management at all stages from the pre-design stage of procurement through to decisions on workspace design, refurbishment and retrofits.

Some of the areas in which facility managers may be involved and which affect building value include the following.

20.4.1 Worker productivity

Spatial layout and indoor environmental conditions are major determinants of the productivity levels of workers who occupy buildings. Facility managers through their involvement in managing and maintaining workspaces can provide designers with a great deal of first-hand knowledge of worker comfort and satisfaction, absentee rates, problem relating to air quality, thermal comfort and ergonomic concerns.

20.4.2 Running and occupancy costs

It is generally the facility manager who has the task of controlling and hopefully reducing the ongoing costs associated with operating and occupying buildings. Their experience in this area naturally provides them with direct knowledge of how costs are incurred and what measures can be taken that will minimize such costs. There are now many documented cases which demonstrate clearly that money spent on installing more effective and efficient components (such as low energy light fittings, control systems, variable speed fans and the like) produces ongoing reductions in recurrent costs such as electricity supply. In some instances, particularly those related to the improvement of lighting in offices and similar work environments, these upgrades also produce measurable improvements in worker performance both in terms of volume of output and in efficiency and accuracy.

20.4.3 Procurement methods

Facility managers are regularly involved in procuring all manner of building components and services. Many parts of a building, such as carpets, mechanical plant and furniture require periodic replacement while others, such as lighting fixtures, control systems and data services may be upgraded before their serviceable life is over as improved technologies become available. The methods used for procuring these items are changing with various innovative systems emerging, such as performance contracting, contracts for the purchase of goods which include an ongoing agreement for maintenance and perhaps removal of the goods at the end of a specified period, and contracts for supply which include future replacement and disposal or recycling of goods. The experience gained in the procurement of these items during the operational life of a building is just as valuable in the early stages of a project and the facility manager can therefore provide advice on these matters to the project team at the outset.

20.4.4 Integrated design

It is generally accepted that as much as 80% of the cost of a building is determined in the earliest stages of the design process, and that design and construction may account for less than 1% of the total cost of a building over its lifetime. Modern buildings are very complex artefacts and require a large and diverse team of professional people to design and construct them, yet it has often been the case that a relatively small number of individuals have been responsible for the early stages of the process during which so much of the future costs are committed.

This scenario leaves little opportunity for other interested parties, such as the facility managers who will ultimately have to deal with the shortcomings of the completed buildings when operational, to make any worthwhile contribution to these early decisions, yet it is during these early stages of the process that the greatest influence on the eventual value of the building may be made. The value of an integrated team approach to building design has been demonstrated in numerous widely reported case studies, for example the ING (formerly NMB) Bank headquarters in Amsterdam (Vale and Vale, 1991). Facility managers should be included in these integrated teams so that their practical knowledge of how buildings function at various levels (i.e., technically, as physical mechanisms with many interdependent parts and functions, as business tools which support the productive activities of the entity occupying them, and functionally, as places to accommodate a range of human activities) can be harnessed and so assist in the production of better buildings.

20.5 Conclusion

Value in buildings is neither easily identified nor easily measured yet it is the goal of everyone who builds. The relationship between facility management and value in buildings is, however, more easily definable and has at its heart the experience of facility managers who are more familiar than any other person with the day-to-day realities of how buildings work, what makes some buildings more successful than others, what are the costs that are associated with owning and maintaining buildings and ultimately how a building represents good value for its owners and/or occupants. It is common sense that the experience of these individuals should be brought to bear across the entire process of building procurement, and not ignored until decisions have been made that then hamper them in achieving value for money for many years after the completion of construction. If clients are to gain the best value for their money it is clear that they should involve those who will be charged with the responsibility of making the building work in its lifetime from the outset.

References and bibliography

Bailey, S. (1990). *Office*. Butterworth Architecture.

Becker, F. and Steele, F. (1995). *Workplace by Design: Mapping the High-performance Workscape*. Jossey-Bass.

Best, R. (1999). 'Integrated Design'. In *Building in Value: Pre-design Issues* (R. Best and G. de Valence, eds). Arnold.

Duffy, F. (1998). 'The New Office'. *Facilities Design and Management*, 17(8), pp. 76-79.

Duffy, F., Laing, A. and Crisp, V. (1993). *The Responsible Workplace*. Butterworth Architecture.

Kleeman, W. B. (1991). *Interior Design of the Electronic Office: The Comfort and Productivity Payoff*. Van Nostrand Reinhold.

Langston, C. (1999). 'The Capital Cost Fallacy'. In *Building in Value: Pre-design Issues* (R. Best and G. de Valence, eds). Arnold.

Lovins, A. and Browning, W. (1992). 'Green Architecture: Vaulting the Barriers'. *Architectural Record*, December.

Romm, J. and Browning, W. (1994). *Greening the Building and the Bottom Line*. Rocky Mountain Institute, Snowmass, Colorado.

Smith, P. (1999). 'Occupancy Cost Analysis'. In *Building in Value: Pre-design Issues* (R. Best, and G. de Valence, eds). Arnold.

Turner, G. and Myerson, J. (1998). *New Workspace, New Culture: Office Design as a Catalyst for Change*. Gower Publishing Limited.

Vale, R. and Vale, B. (1991). *Towards a Green Architecture: Six Practical Case Studies*. RIBA Publications.

Zelinsky, M. (1998). *New Workplaces for New Workstyles*. McGraw-Hill.

Multi-criteria analysis

Grace Ding

21.1 Introduction

Ecologically sustainable development is a concern of people from all disciplines. The concept of sustainability in the context of facility management is about minimizing resources and energy consumption, reducing damage to the environment, encouraging reuse and recycling, and maximizing protection to the natural environment. These objectives may be achieved by considering the most efficient option among competing alternatives through the process of project appraisal at an early stage. Often financial return is the only concern in a project's development but the project that exhibits the best financial return is not necessarily the preferred option after consideration of environmental issues.

Since the industrial revolution the world has been predominated by single-dimensional evaluation processes that concern economic efficiency. Cost benefit analysis is the leading tool in this respect and is widely used in both private and public development to aid decision-making. But in reality decision-making is rarely single dimensional. From the end of the 1960s onwards it was gradually recognized that there was a strong need for the incorporation of a multiplicity of conflicting objective functions. There was also an increasing awareness of negative external effects generated from development and the importance of distributional issues in economic development. The usefulness of single-dimensional appraisal techniques in this respect is increasingly controversial (Zeleny, 1982).

The strong tendency to incorporate multiple criteria and multiple objectives in project appraisal has led to a need for more appropriate analytical tools for analysing conflicts between policy objectives. Multi-criteria analysis (MCA) is a technique established in such a way as to provide the required methodology to undertake the evaluation of multiple criteria and objectives in project appraisal situations.

21.2 Principles of multi-criteria analysis

Tabucanon (1988, p. 4) defines MCA as 'the process of decision making in the selection of an act or courses of action from among alternative acts or courses of actions such that it will produce optimal results under some criteria of optimisation'. The aim of MCA is to provide a systematic framework of analysing information with conflicting objectives and criteria so as to make the trade-offs in a complex choice situation more structured. It provides a well-established methodology and operational framework for problem solving.

Carlsson and Kochetkov (1983) indicates that MCA is the area of decision-making that involves multiple attributes, objectives and goals. Attributes refer to the natural characteristics of objects (for example, features, quality) and they are independent from the decision-maker's needs or desires. They are quantifiable by units such as price per m^2, scales such as yes or no and the like. Objectives are closely related to the decision-maker's needs and desires in order to decide which attributes to maximize or minimize. They represent the directions for improvement to bring undesirable situations back to acceptable levels. They significantly influence the identification of alternatives and criteria for project implementation. Goals are definite needs and desires of decision-makers such as maximum productivity and minimum production costs. They refer to a specific target to be achieved or fulfilled.

Attributes, objectives and goals can be extended and referred to as criteria. That is, all criteria are all the attributes, objectives or goals that will be examined by decision-makers during a decision process. Criteria are measures, rules and standards that guide decision-making.

MCA embraces the key elements of any decision problem concerned with multiple attributes, objectives and goals. MCA comes in different forms. Multi-attribute utility theory reflects a preoccupation with how to construct objectives from attributes and evaluates utility functions in terms of the decision-maker's outcome preferences. Multi-objective programming deals mostly with different objectives and does not attempt to regard them as inputs from forming higher-level objective functions. Goal programming reflects concerns with the conditions of attainment by setting goals, target or aspiration levels and is a useful tactic in the pursuit of human objectives.

The purpose of the MCA technique is to determine the degree to which alternatives satisfy each of the objectives and make the necessary trade-offs to arrive at a ranked order that correctly expresses the preferences of the decision-maker.

21.3 The multi-criteria analysis model

Project appraisal may be considered as a continuous process which takes place during the early stage of a development. No matter whether it is a large or small development, there are always many choices during a decision process that have to be assessed and judged. Generally project evaluation goes through several distinctive stages which are interrelated. As indicated in Figure 21.1, the evaluation process for a project using an MCA technique will not be seen as a simple linear process, but rather exhibits a cyclic nature. Each stage can become additional information and form part of the feedback loop to provide further information for a more precise consideration of the previous stage or stages.

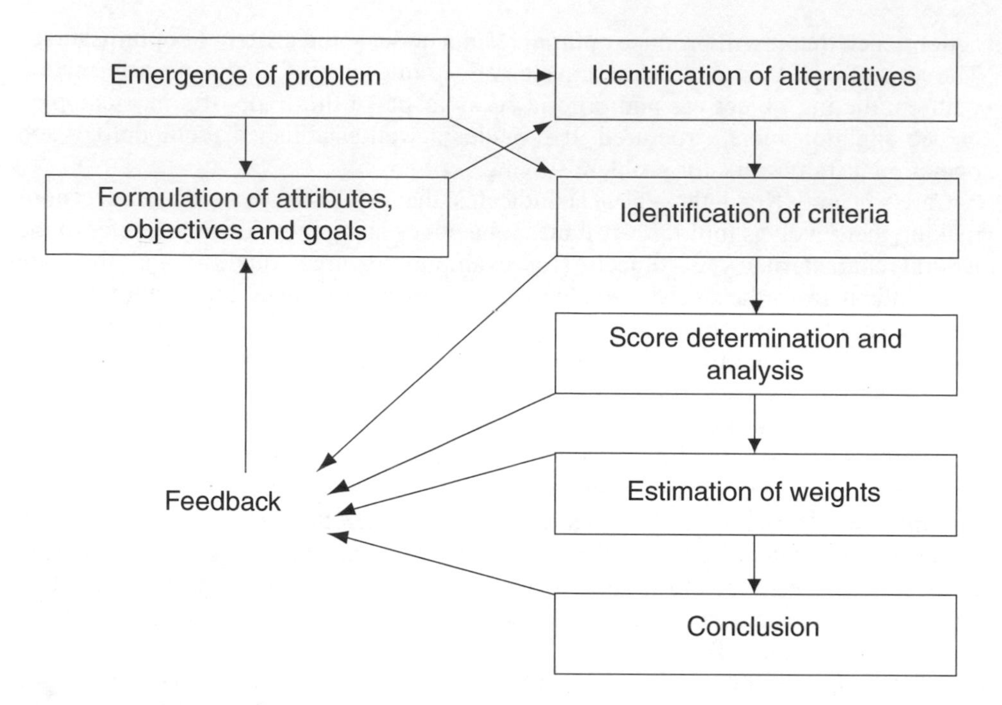

Figure 21.1 Multiple criteria decision-making model

21.3.1 Problem definition

Project appraisal often starts with a definition of a problem and the formulation of project attributes, objectives and goals. A project problem is structured in such a way as to provide adequate specification for objectives and attributes to be identified. Project constraints such as financial, political and external influences will also need to be investigated. Financial constraints relate to the availability of scarce resources for a development. Political constraints have to be considered when public funds are to be utilized. External constraints refer to the effects generated through development upon the man-made and natural environments. These constraints often govern the compilation of alternative and criteria sets in a development. So the earlier the identification of project constraints, the more precise the alternative sets can be arrived at to optimize the best solutions or acceptable compromised solutions to problems.

21.3.2 Identification of alternatives

Based on the structuring of the decision problem, the identification of alternatives is the next step. Alternatives may be in terms of design, location, technology, development options and the like. Alternatives are usually derived from an observation of a problem and through a screening and scoping process. A number of possible options to solve a particular problem may be identified and designed. At this stage, the list of alternatives concerns the objectives of maximizing utilities, optimizing of renewable and non-renewable resources, and minimizing disturbance to the environment. There is no restriction on the number of alternatives but policy-makers tend to reduce the total number in order to facilitate decision-making. It is recommended to have no more than eight alternatives because large numbers of alternatives translates to an increase in uncertainty (van Pelt, 1993).

21.3.3 Identification of criteria

In addition, the evaluation criteria have to be defined. These criteria are a reflection of objectives to be achieved and can be used as a guideline for analysing impacts of individual alternatives. Criteria with respect to environmental effects will also be formulated and this may result in a special environmentally-focused analysis of alternatives at the later stage. The list of criteria should be precise and comprehensive enough to cover the full range of objectives, but again should be limited to no more than eight criteria as this is the maximum number that people can make meaningful and reasonable judgement (van Pelt, 1993). If the number of criteria cannot be reduced to minimum numbers, a hierarchy of criteria may need to be established to separate them into sub-criteria. However, in such a situation the decision process will become more complicated.

21.3.4 Score determination and analysis

Detailed analysis of each criterion is an important step to determine the score in relation to impacts of development. It involves expressing impacts in numeric terms and may be presented in an evaluation matrix with alternatives set against criteria. Each criterion is scored to reflect its relative importance against individual alternative. Scores can be measured in either a quantitative (cardinal) scale or a qualitative (ordinal) scale and the scores may be totalled to show the ranking of alternatives for a project development.

A quantitative scale is expressed in monetary or physical units such as $\$/m^2$, Gj/m^2, etc. However, a qualitative scale is much more difficult to handle and may be expressed in three different scales:

1. Ordinal ranking may be used and it is expressed as 1, 2, 3, … or +++.
2. Nominal scales reflect the characteristics of alternatives such as type of colour.
3. Binary scales contain only two answers such as yes or no.

Whenever a qualitative measure is involved, the measure has to be converted to numerical data (Nijkamp et al., 1990). In the next section the method of transforming qualitative data into quantitative data will be discussed in detail.

21.3.5 Estimation of weights

Weights will be assigned to each criterion to reflect the relative priority of criteria. Nijkamp et al. (1990) defines various types of estimating for criteria weights, broadly divided into two main approaches of direct and indirect estimation. Direct estimation of criteria weights refers to the expression of the relative importance of the objectives or criteria in a direct way through questionnaire surveys. Respondents are asked questions within which their priority statements are conveyed in numerical terms. Direct estimation method techniques come in various forms:

1. *Trade-off method*. The decision-maker is asked directly to place weights to a set of criteria to all pairwise combinations of one criterion with respect to all other criteria.
2. *Rating method*. The decision-maker is asked to distribute a constant number of points to a set of criteria to reflect their level of importance.
3. *Ranking method*. The decision-maker is asked to rank a given set of criteria in the order of their importance.
4. *Seven-points (five-points) scale*. This method helps to transform verbal statements into numerical values.
5. *Paired comparison*. This is similar to the seven-point scale, but the relative importance of criteria is obtained from the comparison of all pairs of criteria on a non-points scale.

However, all these methods run into trouble when the number of objectives becomes large. In such cases, objectives may have to be structured in a hierarchical model to separate them into different levels.

Indirect approaches are based on investigating the actual behaviour of respondents in the past, that is, their past preference statements concerning weights. Weights are obtained through the estimation of actual behaviour in the past, derived from a ranking of alternatives or through an interactive procedure of obtaining weights by direct questioning of the decision-maker and other parties involved. Hypothetical weights may also be used in some particular projects where analysts prepare their own weights to represent the opinion of a specific group in the community, and policy-makers comment accordingly. Each of these approaches has restrictions and limitations in terms of accuracy and cost. Their usefulness strongly depends on the time required and the attitude of respondents.

21.3.6 Making a conclusion

Finally, conclusions can be drawn and decisions made accordingly. Evaluation may be considered as a continuous activity in the planning process. Throughout the whole process, evaluation feedback loops take place in different routes, which aim at providing further information to define alternatives and/or criteria to satisfy the ultimate objectives to be achieved.

21.4 Types of multi-criteria analysis techniques

Multi-dimensional evaluation techniques are broadly divided into two types: discrete and continuous. Discrete methods display a finite set of feasible choice options for project development and their main aim is to provide a basis for classifying multiple criteria and alternatives in an orderly manner for decision-making. These methods often use an evaluation matrix with alternatives set against criteria for analysis. Weights are attached to reflect the relative importance of each criterion. Continuous methods may encompass an infinite set of choice possibilities.

21.4.1 Discrete multi-criteria methods

Discrete multi-criteria methods come in three different types: quantitative (cardinal), qualitative (ordinal) and mixed discrete methods.

In the quantitative discrete method, weighted summation is the most popular approach for handling quantitative data. This is a utility-based method in which scores and weights are summed to indicate the performance of alternatives. However, the technique of standardization may be required to bring different measurement units into a common dimensionless unit. The concordance method is another evaluation method for quantitative data. In an impact matrix, ranking of alternatives are arrived at by means of a pairwise comparison of alternatives. The relationship for each pair of alternatives is derived by establishing both an index of concordance and an index of discordance. The concordance index indicates one alternative is better than the other, whilst the discordance index reflects one alternative performs worse than the other. The higher the concordance index, the more attractive one alternative above the other in the concordance sets. The discordance index is calculated to indicate the maximum difference of scores for alternatives. Alternatives that demonstrate high values for the concordance indices and low values for the discordance indices are preferred.

Qualitative discrete methods comprise two approaches: direct and indirect. In the direct approach, qualitative ordinal data is directly used in the analysis and the results are observed. In the indirect approach, qualitative data is converted into cardinal data first and cardinal evaluation methods are applied thereafter. This approach prevents loss of information but a sufficient basis for cardinalization to carry out the conversion is essential.

There are several types of discrete qualitative multi-criteria methods:

1. The frequency method is a qualitative concordance analysis and is particularly useful for a small number of qualitative categories for both plan impacts and criterion weights.
2. The permutation analysis aims at deriving a rank order of alternatives contained in the evaluation matrix and their attached weight. However, the total number of permutations becomes very large for even moderate numbers of alternatives. This method is most attractive for small numbers of alternatives.
3. The regime analysis can be interpreted as an ordinal generalization of pairwise comparison methods such as concordance analysis.

The final type of discrete multi-criteria method deals with mixed data that is to deal with partly quantitative and partly qualitative information. Evamix is an approach that concerns the construction of two measures: one deals with the ordinal criteria and the other one handles the quantitative criteria. This method results in a complete ranking and provides information on the relative qualities of alternatives.

21.4.2 Continuous multi-criteria methods

The options and actions of decision-makers may be formulated as a continuous range of values. Therefore alternatives formulated in such a manner are often infinite in nature. Compared with the discrete method, the use of pairwise comparison among all elements is not feasible as the continuous method usually deals with an infinite number of alternatives, and alternatives therefore are generally implicit. Multiple criteria linear programming, interactive multi-criteria decision-making and qualitative continuous multi-object approaches are common continuous multi-criteria methods to deal with an infinite set of alternatives.

21.5 Value management

Value management (VM) is defined as 'the identification and elimination of unnecessary product cost', and is a particular representation of MCA commonly used in facility management. The technique concerns optimizing design, function and value, and is not simply cost cutting. Value engineering is a term sometimes used to describe new product design, while value analysis is a term sometimes used to describe existing product improvement.

VM was developed by Lawrence Miles for the US-based General Electric Company. The aim was to improve quality and reduce the cost of materials and labour during World War II. Most of the early VM efforts were aimed at military and industrial hardware applications. The US Navy embraced the technique in the 1950s and initiated the idea of multi-disciplined teams and the VM workshop.

VM in its modern context is a structured, methodical and multi-phased process. It is broken into generic information, speculative, analytical and proposal phases. Sometimes two further phases of implementation and follow-up are also included. Each phase is sequential. The technique operates best in a multi-discipline environment.

Functional analysis is the main activity undertaken in the information phase. Primary (basic) and secondary functions are identified. These functions are related to cost and worth. The worth of the basic function acts as the value standard against which other solutions are compared. Functions need to be expressed in simple terms.

Brainstorming is a common activity in the speculative phase. It consists of a group of people spontaneously producing ideas in an encouraging non-judgemental environment. Research tests conducted at the University of Buffalo (US) demonstrated that groups generate 65% to 93% more ideas than individuals working alone.

Weighted evaluation is the main technique applied in the analytical phase. Alternatives are scored (usually on a scale of 0–5) against a number of weighted performance criteria (usually on a scale of 1–10). The multiplication of score and weight is accumulated across all criteria for each alternative. The highest score indicates the best alternative.

A formal VM programme can increase profits or savings, provide a positive means for future improvements, improve the image of an organization, and improve internal operations and communications. VM is essentially a problem-solving methodology. Cost reduction does not have to be the basic parameter, as saving time, improving aesthetics, eliminating critical materials, etc., can be pursued.

21.5.1 Typical generic approach

VM is an effective tool for systematically optimizing the functional performance of a product or activity. It aims to provide improved value for money, but is not about cost cutting at the expense of quality. It takes into account both objective and subjective issues related to design, construction, operation and management. VM is a comparative decision-making aid.

The information phase is about gathering information. Identification of the function(s) is critical. The primary (basic) function is often expressed as a verb and a noun (for example, the function of a 'chair' is to 'support weight'). Secondary functions may also exist (for example, a chair may also need to be stackable and portable). The cost and worth of the primary function can be identified so areas of poor value are targeted.

The speculative phase is about finding alternative solutions. It is a creative process that uses brainstorming as an idea-generation tool. A facilitator assists the small multi-disciplined VM team to generate ideas without criticism, assessment or negative thinking. New ideas through brainstorming may deliver better value and solve design problems. Innovation can result from lateral thinking.

The analytical phase is about evaluation and ranking. The principal tasks are to evaluate, refine and cost analyse the ideas and to list feasible alternatives in order of descending savings potential. A weighted evaluation matrix is often employed. Each alternative is judged against established performance criteria. Scores are used as the basis for selection.

The proposal phase is about making a clear recommendation. The VM team must thoroughly review the alternative being proposed to ensure that the highest value is being realized. A written report should be prepared to document the recommendation and the processes that were used to arrive at the decision. Life-cost estimates should be attached.

The implementation phase is about ensuring that the recommendations can be enacted. The VM team must present a plan for implementing the proposal, for if management cannot be convinced to implement the change all the work amounts to naught. Since implementation costs can be prohibitive, it is important that VM studies are undertaken early before abortive work occurs.

The follow-up phase is about providing feedback. It is important that there is an opportunity for the VM team to evaluate their work in the light of actual performance. A kind of post-occupancy evaluation needs to be undertaken to determine success. Success ratios and lessons learnt can be documented to indicate whether the VM process was effective and to improve future decisions.

21.5.2 Value management tools

The foremost approach to creativity in VM is the brainstorming technique. A brainstorming session is a problem-solving conference wherein each participant's thinking is stimulated by others in the group. During the session the group is encouraged to generate the maximum number of ideas possible. Multi-disciplined teams are more effective because of the breadth of experience possessed.

A FAST diagram is also a useful tool. FAST is an anagram for Function Analysis System Technique. The approach is used to assist with the analysis of functions. Two questions are used to construct the diagram: 'WHY is the function needed?' and 'HOW is the function performed?'. A network is created where answers to the WHY questions are shown to the left of each function and answers to the HOW questions are shown to the right.

Paired evaluation involves the systematic comparison of pairs of performance criteria to determine which is more important (and perhaps by how much). This approach is used to rank and weight performance criteria for use in a weighted evaluation matrix. It helps to justify decisions and is useful as the number of performance criteria increase.

Weighted evaluation matrices are used to judge alternatives (in rows) against weighted performance criteria (in columns). An index for each alternative is effectively created by multiplying a score depicting performance level against the criteria weight and accumulating these across each alternative. Accuracy for the scores and weights is often best treated as whole numbers rather than decimals.

In the case of the redesign of an existing product or system, 'before' and 'after' sketches can be quite useful. These sketches clearly identify the changes that have been made. The after sketches are used as the basis of the design documentation that is to be assembled and form a benchmark against which the final design can be compared.

VM studies should include an objective assessment of life-cost. Since VM studies are comparison exercises between alternatives or between existing and new solutions, life-costs must be expressed in discounted terms. Comparative costs should include for differential revenue, residual value, tax concessions, etc., over the identified time horizon.

21.5.3 Application in practice

The construction industry generally uses a compartmentalized approach to design. Each specialized sub-group is responsible for issuing, reviewing and updating the criteria and requirements of its own domain. This approach tends to emphasize the performance and costs of the part without due consideration to the performance and costs of the system as a whole.

VM workshops are meetings that are held to identify areas of saving and to make recommendations for improvement. VM workshops are multi-disciplined. Workshops are usually facilitated by a person skilled in getting other people to talk together and to motivate and focus their thinking. VM workshops are often a formal requirement of the conditions of consultant engagement. Some VM workshops might be held over half a day, while others have a longer duration. Workshops can be used during strategic planning, design concept formulation, detailed design, construction and as part of facility management processes. A common approach is to have a 40-hour VM workshop (i.e., one-week duration) at sketch plan stage to help optimize the developing design.

Functional analysis involves clearly identifying what things actually do, or perhaps more importantly, what they must do to achieve the project objectives. Through this process it is possible to identify waste, duplication and unnecessary expenditure and provide opportunity for value to be improved. VM also reviews and tests the assumptions and needs contained in the original project brief.

A VM report should always be prepared to document the outcome of the process. It should contain the following:

1. A brief description of the project studied.
2. A summary of the primary/secondary functions.
3. The result of the functional analysis showing existing and proposed designs or solutions.
4. Technical data supporting the new solutions.
5. Life-cost estimates including risk analysis.
6. A clear recommendation plus a plan for implementation and provision of feedback.

21.5.4 Procedural steps

As the VM process is a highly structured and methodical process, it usually follows a clearly defined procedure. The main steps in this procedure are:

1. *Analyse functions*. Having selected an area of investigation, the first step is to identify the primary and secondary functions. The primary function should be expressed simply in 'high level' terms where possible. Secondary functions are not always relevant. If any primary or secondary function is not satisfied by a proposed solution, then the solution is rejected.
2. *Choose performance criteria*. Performance (assessment) criteria should not be confused with secondary functions. These are the means to be used for differentiation and ranking. Performance criteria are weighted normally using a scale of 1–10 (10 means most important). The paired evaluation technique is useful in ranking criteria and developing their weights. Ignore criteria better measured in dollar terms.
3. *Find alternate solutions*. Using a brainstorming approach, develop a list of possible solutions for delivering identified functions. No judgement or criticism must be allowed to occur until the brainstorming activity is complete. Suggestions that clearly do not satisfy primary and secondary functions are marked as unsatisfactory and dropped. Other solutions are carried forward for analysis.
4. *Analyse performance*. Using a weighted evaluation matrix, each proposed solution is judged against identified performance criteria. The level of performance is

normally scored on a scale of 0–5 (5 means highest performance). Each performance score is multiplied by the criteria weighting and accumulated across all criteria. The higher the total score the better.

5. *Calculate life-costs*. Each proposed solution is also analysed from a life-cost perspective. Capital and operating costs per repeating unit are estimated over the selected time horizon. Results are expressed in discounted terms. A sensitivity analysis is undertaken on the discount rate and perhaps the time horizon. Life-cost is divided by value score to arrive at a value for money index (lowest is best).
6. *Make recommendation*. A recommendation is normally made on the basis of the value for money index, although other issues can be introduced. For example, the value score and life-cost can be weighted so that one has more impact on the final decision than the other. Recommendations should consider implementation and follow-up issues. A written report of the whole process is prepared.

VM is therefore an organized effort applied to the analysis of functions, components, goods and services. It looks at the system as a whole. It ensures required functions are delivered at minimum life-cost without sacrificing the intended level of performance and quality. For examples where the technique has been applied to public infrastructure projects, savings averaging 14% have been realized.

VM applications in the construction/property industry include but are not limited to:

1. Establishing and verifying project objectives.
2. Analysing project briefs.
3. Optimizing design solutions.
4. Resolving conflicts and improving communication.
5. Creating and analysing a range of options for executive consideration.

A VM facilitator is a person skilled in managing people and meetings, and who can guide the VM team towards productive conclusions. The facilitator must ensure that areas of maximum potential are targeted so that savings can be maximized within available time constraints. Facilitators who are experienced in the issues being discussed are preferred.

21.6 Conclusion

MCA can be applied to a range of decision-making problems related to the management of facilities. In particular, it has application to the acquisition of new facilities, and can be used to bring a number of important criteria to bear on decisions rather than limit the investigation to a purely monetary analysis. For example, including criteria such as welfare maximization, energy conservation, waste reduction and other more qualitative issues is possible with an MCA approach.

VM is a specialist form of MCA, and can use many of the processes described for assessing criteria and ranking alternatives. Commonly a paired comparison method is employed to derive weights for criteria, and a weighted matrix evaluation used to score the performance of nominated alternatives. The incorporation of brainstorming within the VM process is a feature of the technique.

Facility managers can benefit from the understanding and use of MCA. The technique can help to fulfil the overall objective of improving quality, reducing cost and minimizing risk.

References and bibliography

Alexander, K. (1996). *Facilities Management: Theory and Practice*. E. & F.N. Spon.

Atkin, B. and Brooks, A. (2000). *Total Facilities Management*. Blackwell Science.

Barrett, P. (1995). *Facilities Management: Towards Best Practice*. Blackwell Science.

Carlsson, C. and Kochetkov, Y. (1983). *Theory and Practice of Multiple Criteria Decision Making*. North Holland.

Cotts, D. G. (1999). *The Facility Management Handbook* (2nd edition). AMACOM.

Ding, G. K. C. (1998). 'The Influence of MCDM in the Assessment of Sustainability in Construction'. In proceedings of 14th International Conference on Multiple Criteria Decision Making, Charlottesville, June.

McGregor, W. and Then, D. (1999). *Facilities Management and the Business of Space*. Arnold.

Moos, R. H. (1996). 'Understanding Environments: The key to improving social processes and program outcomes'. *American Journal of Community Psychology*, 24, pp. 193-201.

Nijkamp, P., Rietveld, P. and Voogd, H. (1990). *Multicriteria Evaluation in Physical Planning*. North-Holland.

OECD (1994). *Project and Policy Appraisal: Integrating Economics and Environment*. Organization for Economic Co-operation and Development.

Park, A. (1994). *Facilities Management: An Explanation*. Macmillan.

Rondeau, E. P., Brown, R. K. and Lapides, P. D. (1995). *Facility Management*. John Wiley & Sons.

Schulze, P. C. (1999). *Measures of Environmental Performance and Ecosystem Condition*. National Academy Press, Washington.

Stevens, D. (1997). *Strategic Thinking: Success Secrets of Big Business Projects*. McGraw-Hill.

Tabucanon, M. T. (1988). *Multiple Criteria Decision Making*. Elsevier.

Tompkins, J. A. (1996). *Facilities Planning* (2nd edition). John Wiley & Sons.

van Pelt, M. J. F. (1993). *Ecological Sustainability and Project Appraisal*. Averbury.

Zeleny, M. (1982). *Multiple Criteria Decision Making*. McGraw-Hill.

Building quality assessment

The quality of an existing facility directly contributes to the operations of the organization and the image it portrays to its customers. Quality relates to both design issues (such as aesthetics, functional layout, flexibility and durability) and operational issues (such as cleanliness, security, fire safety and climate control). It is necessary to be able to assess this quality standard, determine trends over time and benchmark performance against other facilities of similar type. Quality considerations are closely linked to financial ones and therefore attempts to increase quality standards usually come with a price tag.

Quality can be assessed using a weighted evaluation matrix. Similar to techniques used in MCA and VM, a weighted evaluation matrix sets performance criteria, assigns relative proportions to each, and then permits an assessment score to be determined against each criterion. The resultant calculations yield a total score that indicates the merit of the facility as benchmarked against typical results from previous investigations. Standard computer-based models exist for building quality assessment that essentially follow this methodology. One of the most well known and accepted is BREEAM produced by the Building Research Establishment (BRE) in the UK.

Building quality is also important so that occupants are protected and are able to undertake their business in comfort and without hazards. Statutory bodies set out minimum standards of building compliance that must be observed. This is a key responsibility of the facility manager. Failures in providing a suitable environment can not only lead to fines and rectification orders, but also to actions brought about by aggrieved building users. In recent times issues such as passive smoking have led to lengthy court battles and/or large pay-outs by organizations. The link between cooling tower maintenance and Legionnaires' Disease is another example.

Facility managers must ensure that due diligence is exercised. They have a position of responsibility in regard to protecting building occupants who trust that proper measures are taken and regular checks occur. Cases of alleged negligence will arise where agreed procedures have not been followed, where actions are judged as careless or where known problems or hazards are not immediately rectified. Public buildings, in particular, are exposed to significant risk in this regard.

It is normal to conduct regular premises audits for maintenance planning purposes. This is an opportunity to check compliance with standards and by-laws and to schedule any rectification work that may be appropriate. Urgent action must be able to be promptly commissioned, and budgets should plan for this eventuality in advance. There also needs to be a system of follow-up to confirm that the work has been properly completed. Downtime in what may be classified as 'mission critical' services will impact directly on the performance of the business and may lead to substantial financial loss.

The concept of a premises audit can be extended to embrace a review of user satisfaction, commonly known as a post-occupancy evaluation. Rather than perform these now and again, it is a useful strategy to implement a regular process of review and feedback that input into action plans. Satisfaction levels for building employees have a direct correlation to productivity output, motivation and work quality.

As discussed above, the quality of a facility is linked to its performance and those of its inhabitants. This is no more obvious in the area of environmental health and safety (sometimes known as occupational health and safety). Facility managers need to develop and implement health and safety plans and ensure that procedures are communicated to building users holding supervisory positions. Issues include workplace safety, hazard reduction, fire safety, ergonomics, emergency and evacuation procedures, waste disposal, accident reporting and the like. It is all about being diligent in the management of an environment within which others spend a good deal of their time.

This part contains an overview of building quality assessment and the rationale for regular review and benchmarking. Chapter 22 summarizes the emerging field of due diligence and its legal basis. Chapter 23 outlines procedures for undertaking a premises audit, including the role which benchmarking plays. Chapter 24 concludes with an examination of environmental health and safety procedures and the important role of environmental auditing in facility management.

Quality, like value, is hard to define, although we know it when we see it. However, in a professional context, the assessment of quality is not something that should be left to personal opinion, but should be the result of an objective process of review and comparison to what is being achieved elsewhere. Measurements of quality standards can contribute to the value of facilities and their attractiveness to other stakeholders such as tenants and employees. It is worth the effort to ensure that facilities operate at a level conducive to productive work practices. After all, facility management is about ensuring that infrastructure supports core business tasks.

Due diligence

John Twyford

22.1 Introduction

The need to have a due diligence assessment carried out is increasingly a factor of commercial life. The expression 'due diligence' was originally a legal standard that has now become a term of art. Frequently contracts, especially for the supply of services, would require one or both the parties to perform their obligations with due diligence. In the case of a contract for the construction of a building, this meant that the contractor must execute the works with due diligence or face the prospect of having the contract terminated. In *Greater London Council* v. *Cleveland Bridge & Engineering Co Ltd* (1984) 34 BLR 50, Justice Staughton said: 'I consider that [the expression due diligence means] … an obligation on the contractors to execute the works with such diligence and expedition as were reasonably required in order to meet the key dates.' The expression is therefore very much bound up with the notion of timely compliance, with obligations imposed both by contract and the law generally. Due diligence affects the responsibilities of a facility manager in a range of ways including, but not limited to, property acquisition and operational compliance.

A person proposing to acquire property will be very much concerned at the level of compliance attained by the vendor of that property. The compliance here referred to relates to the ongoing obligations that the vendor has been subject to and the obligations that the vendor has undertaken in respect of the transfer of the property. The former may be referred to as 'compliance due diligence' and the latter as 'transactional due diligence' (Duncan and Traves, 1995). The matters disclosed by a due diligence report will have a bearing on the value to the purchaser of the transaction and perhaps a bearing on the negotiation process.

22.2 What is due diligence assessment?

A due diligence assessment is a report prepared by an appropriately qualified expert after investigation for the purposes of acquisition of real or personal property in the context of a commercial transaction. The property might be a business, shares or real property. The nature of the investigation will depend on the property being acquired. With shares it will be necessary to examine the viability of the corporation in question. Where the purchase is of real estate there will be a need to investigate the title and the physical condition of buildings situated on the land. Depending on the nature of the instructions to the person making the assessment, it might be necessary to investigate business and Government records and to establish that the vendor has complied with the law and been truthful in representations made about the transaction. Where a business is being acquired as a going concern it might be necessary to examine the staff to ensure their continued loyalty to the new owner. Where capital assets are being acquired it may be necessary to report on the suitability of those assets for the continued operation of the business in the hands of its new owner. This might involve life-costing of a building.

Needless to say, it is unlikely that one expert would have the expertise to carry out this investigation single-handed. In most cases it will be necessary for the person carrying out the investigation to engage other experts to assist. A fairly obvious example relates to real estate where it would be usual to engage a lawyer to investigate the title to the land. For this reason a team is frequently assembled to prepare the final report. The Australian National Competency Standard for Quantity Surveyors relating to technical due diligence reports lists the following matters as functions of a person carrying out a due diligence report:

1. Establish project objectives and parameters and format of report.
2. Access available data and information.
3. Activate consultant team.
4. Carry out appropriate investigation and prepare initial condition statement including any adverse factors.
5. Provide advice to clients, which outlines potential cost of ownership after analysis of results. (Australian Institute of Quantity Surveyors, 1995)

22.3 Purpose of due diligence assessment

In essence, the commissioning of a due diligence report is part of a risk management process. Since the process is related to the acquisition of an asset, it is incumbent on the purchaser to understand the risks about to be taken. This will enable the purchaser to make appropriate financial provision against the happening of the event constituted by the risk. There will be times when the purchaser will wish to avoid the risk altogether by transferring it to the vendor by contract or insurance. On other occasions the risk will be such that the purchaser will elect not to proceed with the transaction. These things can only happen if the risk is known.

There is an increasing tendency of both public and private sector owners of property portfolios to commission ongoing due diligence assessments to ensure that the value of the asset is monitored as an aid to investment and refurbishment decisions. In addition the assessment will give owners and tenants assurance that buildings are being properly maintained and certified in accordance with regulations and compliance standards. It also allows them to understand where operating procedures can be improved to lower costs without compromising quality, safety or performance.

Christopher Davis in his online paper 'Strategic Issues in Due Diligence' (http://www.davisco.net/ddstrategic.html) makes the point that a purchaser of a business will want the due diligence assessment to show that:

1. The assets that are the subject of the transaction have the value that the vendor assigns to them.
2. The vendor has a good title to the assets.
3. There are no risks that reduce the value of those assets, for example the asset being leased to a third party.
4. There are no other liabilities that may adversely affect the vendor's ability to satisfactorily complete the transaction, for example a taxation liability.

The author notes that a warranty (contractual promise) from the vendor that gives the purchaser assurance on these issues is of limited value only. It should be remembered that in the final analysis a contractual promise is only as good as the person giving it. Warranties suffer the following vices:

1. They last only for a stipulated period of time.
2. A warranty can be used to limit the vendor's liability.
3. Warranties are often disputed.
4. Warranties are difficult to enforce, often involving the need for litigation.

It is possible in some jurisdictions, notably the United States, to insure against breach of warranty.

22.4 The due diligence team

Since a prime purpose of obtaining a due diligence assessment is to make a purchaser of an asset aware of potential liabilities that they might be exposed to as a result of the purchase, it is important to understand what those risks might be. The risks will vary with the nature of the asset. Where the purchase of a commercial enterprise is under consideration the emphasis will be on the economic good sense of the proposal, whereas if it is a real estate asset under consideration issues of title, land use and condition of buildings will be important. With these matters in mind Davis has suggested a number of areas of risk that should be considered. Since the emphasis of this chapter is facilities management, a number of topics have been added to accommodate this aspect of the discussion:

1. *Title to any real estate assets.* It is important for a purchaser to know that a vendor can convey an unencumbered title to the purchaser that will enable the purchaser to use the asset as desired. An unregistered lease or adverse claim to title to the land or part thereof will have an effect on the value of the proposed purchase.
2. *Land use.* In many jurisdictions town planning controls restrict the present and potential use of land. Where an industry is being carried on, a potential purchaser will need to be assured that the land can continue to be used as contem-

plated. In this regard it is important to realize that some land use consents have a sunset clause.

3. *Physical condition of buildings and infrastructure*. A potential purchaser will need to know the condition of the improvements on land and the likely cost of repair, restoration and maintenance.
4. *Appropriate use of assets*. A purchaser will need to know whether the capital assets of an enterprise are being utilized to capacity. It may be more appropriate to lease premises or subcontract some of the functions of the business.
5. *Risks relating to the environment*. Failure to comply with the increasing regulation of environmental matters can lead to civil liability and criminal prosecution. Another factor is that the arm of this branch of the law often reaches to directors and officers of corporations. Where a purchase of a business enterprise is accomplished by the purchase of shares in a corporation, the corporation has a continued legal existence. In the hypothetical example of a manufacturing corporation discharging industrial effluent into a river, it would be the corporation and its present officers who would be prosecuted, the real culprits having long since departed.
6. *Health and safety*. A business enterprise has a potential liability to both its workers and to third parties. It is possible for such an enterprise to have a substantial liability on either of these counts. An Australian corporation recently faced substantial claims for damages to former workers who had been exposed to asbestos particles some thirty years earlier. In terms of liability to third parties, the Union Carbide experience at Bhopal stands as a stark reminder.
7. *Legislative changes*. The law can change in a way that makes a particular enterprise either illegal or less profitable. Although a good deal of foresight and political acumen are needed here, the law is a matter that a purchaser of an asset or enterprise will be interested in. Consumer and competition laws sometimes change in ways that inhibit business activities.
8. *Negligent advice*. It is possible that an enterprise will be suffering from, or has the potential to suffer from, negligent advice. Recently in Australia a motoring organization received advice as to how it might demutualize, which proved incorrect. As a result of the advice the organization wasted A$30m.
9. *Political and social instability*. This is a factor that needs to be taken into account, particularly when acquiring foreign assets.
10. *Product liability*. Claims for product liability are often made many years after the offending product has been sold. A purchaser who acquired the shares in a corporation would also acquire the liabilities. Products that have thus far involved their manufacturers in litigation include cigarettes, IUDs and silicone implants.
11. *Property damage and loss*. The physical state of property needs to be checked and reported on. In addition, ownership of intellectual property needs to be verified and the right of a vendor to pass on a good title confirmed.
12. *Unpaid taxation*. Benjamin Franklin noted that the two certainties in this world were death and taxes. Again, in dealing with a corporation it should be remembered that the corporation's obligation to pay tax remains with the corporation irrespective of changes in ownership of the corporation.
13. *Unpaid suppliers*. The risk posed by a failure to pay suppliers speaks for itself.

The team that carries out a due diligence assessment will be managed by core personnel from the organization responsible for the task. In addition it may be necessary to get advice from any one or a number of the following disciplines: accountancy, merchant banking, risk management, legal, actuarial, environmentalists, health and safety experts, product liability analysts, survey, architectural, engineering and valuation. The reference to merchant banking is on the list given by Davis. It would seem that in the United Kingdom, where this type of assessment is carried out on a more formalized basis, the aid of a merchant banker is enlisted for advice on strategy, especially for the acquisition of a business enterprise.

22.5 The process of due diligence

It is more appropriate to see the provision of a due diligence assessment as part of an ongoing process rather than the mere provision of a report. This is because there will be many times when the person responsible for the report will need to take the instructions of the client and this interaction will determine the course of the transaction. Davis lists a number of steps in a typical due diligence commission in the United Kingdom that is generally reproduced hereunder. The steps are given with the caveat that these transactions appear to follow a more stylized form in the United Kingdom and the steps refer to the sale of a business rather than a real estate asset:

1. Where a merchant bank is involved, some initial due diligence is undertaken, covering such areas as political risk and compatibility of organizational cultures. Davis was directing his remarks where one corporation was proposing to take over a foreign corporation.
2. The purchaser may undertake some basic due diligence, usually by way of oral discussion with the vendor. The purchaser may also involve its accountants in some preliminary analysis of the vendor's final accounts. Here the commonsense initiation steps for any commercial transaction are subsumed into the rubric of due diligence.
3. The purchaser negotiates basic heads of terms with the vendor.
4. The purchaser instructs its accountants to commence due diligence. Note that in many transactions it would not be desirable or necessary to regard a due diligence assessment as an extension of an accounting function.
5. The purchasers and vendors will instruct their lawyers to begin preparing the necessary documents. The purchaser's lawyers will forward a set of preliminary enquiries to the vendor's lawyers. In response to this the vendor will usually warrant the accuracy and completeness of the written responses to those enquiries.
6. The vendor's lawyers will pass the preliminary enquiries on to the vendor's nominee for handling enquiries, who will arrange for the requested documentation to be collated and the questions answered. These will be returned to the vendor's lawyers who will vet the responses and filter them where considered appropriate, before restating the responses in their own terminology and forwarding them to the purchaser's lawyers. These documents are indexed and referred to as the disclosure bundle.

7. Gaps in the written responses by the vendor's lawyers and the disclosure bundles are filled from time to time by written responses and the provision of further documents.
8. The purchaser's due diligence team prepares their due diligence report.
9. The purchaser's accountants prepare their draft due diligence report.
10. The purchaser's solicitors may provide a due diligence report.
11. The negotiations concerning the warranties are finalized close to the exchange of the sale agreement. It is pointed out that at the exchange of the sale agreement the parties are bound to go ahead with the transaction. The duc diligence process is directed to finding if the purchaser should go ahead with the transaction at all, and if so what warranties would be required from the vendor. The requirement for warranties could arise out of matters brought to light by the due diligence assessment.
12. The vendor's lawyers will then produce a draft disclosure letter disclosing fact and documents against the warranties. This would be where the warranty required the vendor to produce proof of a particular matter, for instance, the existence of a town planning consent could be satisfied by production of the certificate from the appropriate authority. Other warranties will require the vendor to make contractual promises about aspects of the transaction. An example would be a warranty that the income was above a certain level. If the matters required by the warranty are not fulfilled the purchaser will have grounds for legal action.

22.6 Reporting

It is clear that a person who takes the trouble to commission a due diligence assessment report is likely to order their affairs on the basis of the content of the report. For this reason the report must be accurate, as a failure in this regard will lead to the person who was responsible for the report being sued. It is suggested that a good report will exhibit the following characteristics:

1. The author of the report should state their qualifications to make the report.
2. It should be in writing, containing a clear statement of what were the instructions to the consultant. The report should answer all of the questions posed, but only those questions.
3. As the author will almost certainly have been present when the scope of the report was determined with the client, it is important that the report canvasses all issues that are needed for the client's protection. A failure to include a relevant area of investigation might of itself be negligent.
4. The report should differentiate clearly between matters of observation and matters of opinion.
5. Where a matter of opinion is given, it should be related to observed facts or the opinion of other experts.
6. Where the opinions of other experts are included in the report, they should be clearly identified as such.
7. Where the author has a reservation as to an issue on which they are reporting, this

reservation should be clearly identified and where appropriate further investigation advised. For example, it might not be possible to gain access to a certain part of a property to inspect it.

8. The temptation to 'guild the lily' or tell the client what it is thought he wants to hear should be resisted.

22.7 Extent of legal liability

Invariably the relationship between the person preparing the report (expert) and the client is one of contract. For a consideration, the expert agrees to carry out professional services. The degree of formality of the contract will depend on the particular transaction. A person who prepares a due diligence assessment report could be in breach of their contract to do so and guilty of negligence.

The tort or civil wrong of negligence is by legal standards a relatively recent phenomenon. Originally, only those persons carrying on what was described as a common calling could be held accountable in negligence. This was on the basis that innkeepers, carriers, surgeons, attorneys and the like held themselves out to the public as having certain skills and if these skills were not displayed the person was liable for the loss caused. There was no general incidence of liability applicable to all members of society.

The advent of negligence as a discrete tort came as a result of a famous snail in a bottle of ginger beer. In 1932 the *House of Lords in Donoghue* v. *Stevenson* [1932] AC 562 enunciated the principles that were to become the basis of liability in negligence. The plaintiff became ill after consuming ginger beer contaminated by a decomposing snail. The defendant was the manufacturer of the ginger beer who supplied the product in an opaque bottle thereby precluding inspection before consumption. Since the plaintiff was not the purchaser she could not sue in contract. Lord Atkin said:

> 'At present I am content myself with pointing out that in English law, there must be, and is, some general conception of relations giving rise to a duty of care, of which the particular cases found in the books are but instances ... But acts or omissions which any moral code would censure cannot in a practical world be treated so as to give a right to every person injured by them to demand relief. In this way rules arise which limit the range of complainants and the extent of their remedy. The rule that you are to love your neighbour becomes in law, you must not injure your neighbour … Who, then, is my neighbour? The answer seems to be persons who are so closely and directly affected by my act that I ought reasonably to have them in my contemplation as being so affected when I am directing my mind to the acts or omissions which are called in question [the duty of care].'

The subsequent history of the tort of negligence has essentially been a description of how the duty of care has been extended to new situations. In *Hedley Byrne & Co* v. *Heller & Partners Ltd* [1964] AC 465, a merchant bank was held to owe a duty of care in the giving of a credit reference. Lord Reid said:

> 'A reasonable man, knowing that he is being trusted or that his skill and judgment were being relied on, would, I think, have three courses open to him. He could keep silent or decline to give the information or advice sought; or he could give the information with a clear qualification that he accepted no responsibility for it or that it was given without that reflection or inquiry which a careful answer would require; or he could simply answer without any such qualification. If he chooses to adopt the last course, he must, I think be held to have accepted some responsibility for his answer being given carefully, or to have accepted a relationship with the inquirer which requires him to exercise such care as the circumstances require.'

The cases where a plaintiff has succeeded against a defendant who has given advice, supplied information or rendered professional services have multiplied. A solicitor who had failed to notify an executor of the death of the testator thereby delaying an application for probate and causing loss to the estate was held to be negligent (*Hawkins* v. *Clayton and others* [1988] 164 CLR 539). The duty of care was said to arise from the relationship of proximity between the executor and the firm of solicitors.

Perhaps a note of reassurance that might be added is that, although a duty of care exists, the liability is not absolute. The standard of performance to be expected is that of a reasonably competent member of the calling in question. In *Voli* v. *Inglewood Shire* [1963] 110 CLR 74, an injured visitor to a building successfully sued the design architect after the building collapsed. Of the standard of care as opposed to the duty (which was held to exist), Justice Windeyer said:

> 'He [the architect] is not required to have an extraordinary degree of skill or the highest professional attainment. But he must bring to the task he undertakes the competence and skill that is usual among architects practising their profession.'

Similarly, it was said that an auditor 'is not an insurer' or 'detective' (*Saif Ali* v. *Sydney Mitchell & Co* [1980] AC 198). A person who has suffered at the hands of an incompetent professional would seem to enjoy rights similar to those described above under Article 1382 of the French Civil Code or Article 823 of the German Civil Code.

It is fairly clear that a person who prepares a due diligence assessment for another is under a duty of care to see that the report is carried out with appropriate skill and care. The fact that the expert knows that the report is required to enable the client to make an informed decision about a proposed commercial transaction provides the necessary degree of proximity. The liability for negligence would extend to the selection of an unsuitable secondary expert. Extreme care needs to be taken and appropriate professional indemnity insurance is essential.

22.8 Conclusion

Due diligence is an emerging area of professional facility management practice, particularly in the context of facility compliance with statutory regulations and codes.

The facility manager will often have need to seek the preparation of due diligence reports on a variety of matters to minimize the risk of possible action or non-compliance penalties, and these may be undertaken in conjunction with a detailed physical premises audit. The introduction of due diligence processes to the area of environmental health and safety is also a growing area of professional activity worldwide.

References and bibliography

Australian Institute of Quantity Surveyors (1995). *National Competency Standards for Quantity Surveyors/Construction Economists*. AIQS, Canberra.

Davis, C. (date unknown). 'Strategic Issues in Due Diligence'. Davis and Co. <http://www.davisco.net/ddstrategic.html>

Duncan, W. D. and Traves, S. J. (1995). *Due Diligence*. LBC Information Services.

Premises audits

23.1 Introduction

One way to improve the quality of facilities is to undertake a review of existing performance. This can take many forms. For example, one year after commissioning, a post-occupancy evaluation may be performed to test the completed design against user expectations. An energy audit can help identify areas where energy usage is unnecessarily high or where equipment is operating at a sub-optimal level. An environmental audit may include energy, but also will cover a range of resource issues including recycling, purchasing, waste disposal, pollution, training programmes, etc. A due diligence report looks at compliance issues and recommends facility upgrade where necessary.

Within this collection of review processes, the premises audit (or facilities audit) has special application. Bernard Williams Associates (1999, p. 12-1) describe a premises audit as 'an independent review of the costs of providing and operating premises and facilities, together with performance levels, management structures and systems'. It incorporates gathering and analysing operating cost data and other information that can assist in finding improvements to existing practice, and as such can be undertaken at any time.

Today, facility management is under scrutiny from efficiency experts commissioned to provide cost-effective solutions to an activity that has traditionally been perceived as a necessary evil – a cost of doing business. Along with this analysis is the continuous request that facility managers provide better customer service, improve space comfort, supply reliable mechanical and electrical energy performance, and do so at no more cost than last year's operating budget (McKew, 1996).

23.2 The auditing process

The auditing process is normally undertaken by an external consultant, and therefore enables an independent review to occur. External consultants bring with them a wealth of experience of similar activities from work with other organizations. In this way areas of poor performance are more readily identified.

Bernard Williams Associates (1999) recommends a two-phase process. The first phase involves analysing the organization's own cost data and expressing it in a form that will support external comparison. In-house cost data is typically in a unique and accounting-like format, and much information has to be uncovered to make sense of it and to understand its context. The analysis will express costs into categories (cost centres) per square metre of floor space or in some cases per employee. The purpose of the first phase is to target areas that require further study. This can be achieved by the application of benchmark data gleaned from past studies.

The second phase investigates any anomalies uncovered in the first phase and recommends corrective action. The establishment of performance indicators can help to measure the success of corrective action. While the main focus may be operating costs, it can be expanded to cover other performance measures like energy, churn, outsourcing and productivity.

23.3 Cost centres

Cost centres are reporting categories that, if standardized across organizations, can enable effective comparison and benchmarking to take place. Unfortunately, the only consistency found in practice is generally within individual consultant organizations. Bernard Williams Associates (1999) propose a classification system as follows:

1. *Services maintenance*. This cost centre comprises mechanical services, electrical services, plumbing, lifts and escalators and the like.
2. *Building maintenance*. This includes repair on the building fabric, decorations, fittings and external works.
3. *Cleaning*. This relates to internal areas (including amenities), windows and cladding and external areas.
4. *Utilities*. This comprises service charges for electricity, gas, oil, water and sewerage and waste disposal.
5. *Security*. This includes security personnel, service contracts, equipment and associated maintenance.
6. *Special*. This category is reserved for specialist services like landscaping and pest control, and for the provision of supplies and sundry items.
7. *Management*. This covers salaries, benefits, overheads, senior management and general equipment costs.
8. *Improvements and adaptions*. This cost centre includes churn costs, extensions and improvements to existing facilities.

Langston (1991) alternatively proposes a standard elemental classification. Although built on the Australian National Public Works Conference (NPWC) *List of Elements and Sub-elements*, it is reasonably similar to other elemental classifications used in the UK, US and most Commonwealth countries. Within each element (such as Roof, External Walls, Heating and Cooling, Fire Protection, etc.), costs are categorized into cleaning, energy, maintenance, replacement and other expenses. Elements can even be broken down into standard sub-elements for further comparison.

An elemental approach has significant advantages in that it ties into design cost planning processes. The analysis of operating costs is dependent upon component size, such as square metres of carpet or lineal metres of stormwater pipe. Therefore creating a clear relationship between capital works and premises occupancy will enable operating costs to be properly assessed.

23.4 Historical data

Data collected from different facilities will reflect the specific nature of the building, its location and occupancy profile. When measured overall, no two buildings will have identical running costs, nor will the running costs for any specific building be the same from one year to another. Therefore obtained results should be applied to other situations with caution.

Problems with data reuse across facilities stem from the variable durability of materials and systems and the effects of their interaction, the environment and users of the building. This combination of effects will lead to a wide range of results in the data. Long time frames for building component replacement also result in an incomplete picture. Furthermore, the inability of building owners to compile data in an appropriate form inhibits its reuse. Even if all these problems were overcome, the quality of the original construction directly affects operating performance, and therefore the whole issue of workmanship and quality assurance is invoked.

There is a general absence of reliable operating cost data for buildings. The failure to collect suitable data is closely linked with the low level of life-cost planning currently evident in practice. Many argue that without historical data, life-cost planning cannot take place, but others might say that without life-cost planning activities the framework for collection of data and analysis is missing.

The complexity of modern facilities and the interrelationship between systems and components create problems for comparative analysis. Actual performance is highly context-dependent. For example, harsh environments will accelerate deterioration by comparison to benign environments. Occupancy patterns and hours of operation will also affect performance.

The long time frame involved in the measurement of component lives and maintenance intervals makes accurate record-keeping vital. For example, the life of floor covering may be fifteen years or more, and if records are not clear about when it was last replaced it will be difficult to determine its actual life. The changing time value of money also poses a problem for the interpretation of costs over many years.

The inability of organizations to compile data in an appropriate and available form for auditors inhibits its reuse. Such an activity is not normally seen as necessary to the accounting process, and items like dollar value, contractor and date of payment is often all that is recorded. But for the data to be useful, it must contain quantities of work completed, reasons for the work and which items it relates to in the master cost plan.

The quality of the original construction directly affects performance. Expensive and durable systems can be rendered bad investments because they were not installed correctly. Similarly, failure of an inexpensive item can result in damage to an adjoining expensive item. Quality assurance procedures are therefore vital to ensure that original work is of appropriate standard. Where maintenance or replacement is the result of poor workmanship, this should be clearly noted.

The need for vast amounts of historical data across a range of building types over many years dictates that the method of recording be computer-based. An elemental approach is useful (particularly as it will match design cost planning processes), but attached to this classification should be information about context. Electronic conversion of costs to selected base dates and criteria-based search capabilities are useful.

An elemental or sub-elemental approach is therefore preferred as the basis of any classification system for operating costs. This requires consistent standards to be applied across facilities so that data is comparable. Until an adequate source of data is available, or even after it is available, estimation of operating costs from first principles will be more reliable than averaged historical performance. In addition, data derived from the particular facility being audited will be more valuable than data derived from an average of other similar yet unknown facilities external to the organization.

23.5 Monitoring performance

The monitoring and management of actual operating performance is also known as life-cost analysis. Monitoring comprises the recording and categorizing of expenditure incurred from year to year as an input to future decisions. Management comprises the interpretation and action taken to overcome areas of cost over-run or where planned targets have not been adequately realized.

In the context of operating performance, each year costs need to be recorded in current dollars (i.e., actual cost) and converted back to base value for comparison to targets shown in the master cost plan. This is performed using inflation indices. It is important that the records include notes to give a complete picture of the reasons behind the costs.

Although a particular piece of equipment or other component may have remaining life, monitoring of performance may indicate it is operating poorly by comparison with design expectations or other benchmarks. Decisions can be made to terminate an asset prematurely if the introduction of a new solution will deliver savings within a given time horizon. This is a process that gives rise to proposed management initiatives for ongoing improvement.

One of the most important features of monitoring actual performance is the knowledge that is gained about future design decisions. Feedback to designers means that past mistakes are not repeated and successful solutions are identified and built into future specifications. This implies that designers take some interest in their projects after commissioning, and is formally achieved using a post-occupancy evaluation process after one year of operation.

23.6 Benchmarking

Benchmarking is the comparison of performance to the best of industry practice (best-of-breed), with a view to improving performance to match or exceed what others have been able to achieve. Webster's dictionary defines a benchmark as 'a point of reference from which measurements may be made; something that serves as a standard by which others may be measured or judged'. More specifically for facility managers, benchmarking is a systematic process used to identify, understand and adapt practices to improve performance and efficiency.

Benchmarking can come from published data, such as that available by professional associations and service providers, specific data from previous premises audits and peer group consultation (Bernard Williams Associates, 1999). Benchmarking data can fulfil a number of purposes, including:

1. The establishment of a comparative data set from which running costs can be compared both between and within organizations.
2. To increase an organization's awareness of the importance of facility costs in their quest for increased business performance.
3. To establish a base line from which efficiency expectations can be derived.
4. To increase the awareness of the processes that underlie facility costs and performance.
5. To encourage co-operation between private and public organizations so as to better manage the usage of scarce resources.
6. To create industry focus and awareness, which leads to greater emphasis being placed on the facility management industry as both an employment generator and a valuable participant in the national economy.
7. To encourage research leading to further enhancements of the benchmarking process, which in turn will further improve the efficiency of the industry (FMA Australia, 1999).

The context within which facilities operate can be so diverse as to enable the application of published benchmarks to be highly questionable. Furthermore, a focus on cost perhaps belies the real potential of a premises audit to add value to an organization.

23.7 Performance indicators

Performance indicators are often expressed in monetary terms per square metre of floor area or per employee. They are in fact ratios that can be compared between operations of different size and configuration. An example of a performance indicator is operating cost/m^2 of gross floor area. These are useful to compare against industry benchmarks or averages, but overlook the actual context within which the costs are generated.

Operating costs are of importance, but in the context of total turnover are usually quite minor. Bernard Williams Associates (1999) identifies human resource costs as typically 75% of turnover, assets and equipment as 20% and operating costs (including rent and service charges) as rarely more than 5%. In this context, a premises audit aimed at analysing operating costs alone is not that significant to an organization's bottom line.

Alternatively, performance indicators that look at productivity are highly valuable. It is accepted that the performance of a facility directly affects productivity, so it is logical that facilities can be evaluated using productivity as the unit of measurement. While it is unlikely that productivity benchmarks across organizations will be useful (or even available), comparison within an organization is certainly possible. The performance of facilities and the justification for budget allocations and improvements can therefore be assessed against increases in worker productivity. The premises audit can investigate those aspects that are shown to have a negative effect on productivity, with a view to recommending change.

23.8 Value management

The value management (VM) process is a useful tool to explore methods of improving productivity. This is because VM is a decision-making tool that combines both objective (tangible) and subjective (intangible) criteria into a single assessment process. It also has the advantage that it is a consultative team-based procedure, drawing on expertise from a diverse range of people within and even external to an organization.

Furthermore, VM can effectively combine cost and function. Using a multi-criteria approach a value score can be determined for all non-monetary criteria, where each criteria is weighted according to its relative importance. The value score can be divided by the calculated life-cost (capital plus operating cost) per square metre to produce a comparative value for money ratio. The higher the ratio, the better value an alternative delivers. The value score can incorporate a wide range of issues, such as aesthetics and image, productivity enhancement, environmental performance, flexibility and the like.

VM can draw on the outcomes of the first phase of a premises audit and play a major role in the second phase. Anomalies uncovered need to be assessed in context, and solutions to problems are not necessarily obvious or simple. The brainstorming and multi-disciplinary nature of VM provides an excellent forum to find innovative solutions that can add value to an organization.

23.9 Implementing change

It is the responsibility of the facility manager to constantly look for ways of improving the performance of facilities, and to make recommendations to senior management as to the wise use of funds in this regard. With improvements to worker productivity as the goal, the consideration of change is placed in its correct context, breaking the perception that facility costs are a necessary business overhead.

Recommendations arising from a collaborative VM process are more likely to be favourably treated by senior management than those developed in isolation, and have greater chance of acceptance across all sections of the organization, particularly where key representatives were part of the process. The difference, however, between good ideas and valuable solutions is the ability of the idea to be implemented. This dictates an effective written proposal supported by verified facts and consultation. The premises audit takes on increased importance in this regard because it demonstrates areas of sub-optimal performance compared to the 'best-of-breed'.

References and bibliography

Alexander, K. (1996). *Facilities Management: Theory and Practice*. E. & F.N. Spon.

Atkin, B. and Brooks, A. (2000). *Total Facilities Management*. Blackwell Science.

Barrett, P. (1995). *Facilities Management: Towards Best Practice*. Blackwell Science.

Bernard Williams Associates (1999). *Facilities Economics*. Building Economics Bureau Limited.

Cotts, D. G. (1999). *The Facility Management Handbook* (2nd edition). AMACOM.

FMA Australia (1999). *Facility Operating Cost Benchmarks*. Hassel Hunt & Moore, Sydney.

Green, G. M. and Baker, F. (1991). *Work, Health, and Productivity*. Oxford University Press.

Langston, C. A. (1991). *Life-Cost Procedures Manual*. Department of Public Works, Sydney.

McGregor, W. and Then, D. (1999). *Facilities Management and the Business of Space*. Arnold.

McKew, H. J. (1996). 'A Maintenance Improvement Plan: The complete remedy for correcting facilities management deficiencies lies within a comprehensive departmental audit'. *AFE Facilities Engineering Journal*.
<http://www.facilitiesnet.com/fn/NS/NS3afe4.html>

Moos, R. H. (1996). 'Understanding Environments: The key to improving social processes and program outcomes'. *American Journal of Community Psychology*, 24, pp. 193-201.

Park, A. (1994). *Facilities Management: An Explanation*. Macmillan.

Rondeau, E. P., Brown, R. K. and Lapides, P. D. (1995). *Facility Management*. John Wiley & Sons.

Salvendy, G. (1997). *Handbook of Human Factors and Ergonomics* (2nd edition). John Wiley & Sons.

Schulze, P. C. (1999). *Measures of Environmental Performance and Ecosystem Condition*. National Academy Press, Washington.

Tompkins, J. A. (1996). *Facilities Planning* (2nd edition). John Wiley & Sons.

Environmental health and safety

24.1 Introduction

Productivity has been cited as an area of common ground between human resource management and facility management. In the same way, environmental health and safety is a shared responsibility. This gives strength to the argument that the human resource and facility functions can be effectively combined. Nevertheless there is a direct relationship between productivity and environmental health and safety, as problems caused by the latter will reduce the organization's capacity to be efficient and to prosper.

Environmental health and safety encompasses issues of employee well-being, as well as the well-being of customers and other facility users. Common examples include fire safety, accident prevention, lifting and moving heavy objects, hazard prevention and emergency response, ergonomics, sickness and ill-health caused by materials, passive smoking, high noise levels, indoor air quality, lighting and temperature control, pollution, security and the training programmes that ensure that these issues are properly understood and practised.

Government regulation and control exists to protect people and to set appropriate standards of compliance. These vary from country to country, but there is a clear international requirement for facility managers to exercise considerable vigilance about environmental health and safety. Failure to do so may lead to expensive legal action being taken against the organization.

24.2 Environmental management

Environmental health and safety is a component of corporate environmental management. Bernard Williams Associates (1999) identifies health and safety as being primarily an exercise in risk management. Environmental management additionally includes issues of environmental protection, which perhaps indirectly affect health and safety, such as energy conservation and the efficient use of resources.

Environmental auditing is a technique often applied to assess the quality of environmental management practices.

An important debate is the balance between regulatory (mandatory) control and voluntary control. While it is clear that it is in an organization's interest to have effective environmental management, not all organizations achieve this, and therefore government intervention is common.

Systematic environmental management evaluation is critical. International standards for Environmental Auditing (ISO 14010, 14011 and 14012) and Environmental Performance Evaluation (ISO 14031) are intended to be programmatic in nature and to provide a direction for the evaluation of environmental systems and the management at a single facility or for a corporate wide system of facilities.

24.3 Health and safety plan

Jones (1997) believes that there is virtually no area in the workplace today that is immune from regulatory oversight, and no room for claims of ignorance when it comes to employee safety matters. How organizations deal with these issues is of paramount importance to their bottom line in terms of decreased costs for workers' compensation and insurance rates, as well as potential liability from litigation or regulatory fines. A workplace with a written health and safety plan is not only safer, but having such a plan can reduce an employer's liability in regard to local, state and federal laws and regulations. A properly customized and implemented health and safety plan should do the following:

1. Establish a blueprint for health and safety initiatives.
2. Allow managers to identify and eliminate workplace hazards.
3. Demonstrate good-faith efforts in meeting health and safety responsibilities.
4. Provide a framework for employers and employees to work together toward the common goal of safety.
5. Provide the ability to co-ordinate business policies and procedures with industry standards and practices (Anonymous, 1996).

24.4 Important issues

To achieve an adequate health and safety plan, any organizational programme must encompass issues of management leadership and employee participation, workplace analysis, accident and record analysis, hazard prevention and control, emergency response, and health and safety training (Jones, 1997).

24.4.1 Management leadership and employee participation

The health and safety plan starts with top management involvement and includes:

1. Appointing a champion.
2. Continual hazard analysis and job safety analysis.
3. Methods to control hazards.
4. Management, supervisor and worker training and follow-up (Westerkamp, 1995).

Top management can demonstrate its commitment by posting a safety policy and meeting with employees to explain it, visiting work areas and reviewing accident reports, wearing personal protective equipment even when briefly visiting work sites, organizing employees to use their special workplace knowledge as members of the safety committee, and delegating authority and providing sufficient resources for an effective programme.

Health and safety require an ethos in the workplace, fostered by management and enacted by employees.

24.4.2 Workplace analysis

Facility managers should be aware of health and safety legislation regarding business premises and the workplace. Breach of health and safety laws is a criminal offence. In addition to criminal proceedings, employers may find themselves with proceedings in the civil courts for breach of duty of care to employees.

Some of the main pieces of legislation of which those responsible for business premises should be aware include those dealing with all aspects of people at work, including:

1. Fire precautions and welfare.
2. Health and safety at work that imposes duties on all persons at work, and affects all work activities and all work areas. It requires the employer to safeguard so far as is reasonably practicable the health, safety and welfare at work of all employees, and requires the employee to take reasonable care for their own health, safety and welfare and of other persons who may be affected by his or her acts or omissions at work and to co-operate fully with the employer or any other persons to comply with their statutory duties.
3. Protection of employees against substances hazardous to health.
4. Exposure levels for noise at work, electricity management, wearing of head protection on construction sites, etc.
5. Environmental protection that sets out a system of pollution control.
6. Risk assessment to reduce and/or manage the risks.
7. The assessment, provision, training and use of protective equipment (Davies, 1996).

An assessment of risks in the workplace is important. Areas of potential risk or areas that are non-compliant must be rectified immediately. Workplace analysis is an ongoing responsibility.

24.4.3 Accident and record analysis

Zenker (1995) cites accident prevention as a primary focus for any company. A sound accident prevention system saves time and money, and therefore safety can add value to an organization by improving overall productivity and adding to the bottom line. One way facility managers can prove this is by showing management the financial benefits that result from such a programme. The obvious solution to the workers' compensation cost crisis is a proactive accident prevention plan incorporating the factor of behaviour. But as Zenker (1995) explains, facility managers will truly be successful when accountability for workers' compensation costs are transferred to the business unit where those costs are incurred and are factored into the costs of doing business just as the costs of safety are done currently. Accident prevention through a behavioural safety process with high employee involvement is the key to achieving reduction of workers' compensation costs.

A safe, healthy environment for employees and users of educational, healthcare, commercial and government facilities depends directly on ensuring that health and safety plans are applicable to conditions. A careful review of records and historical safety experience for the facility and others like it is important. This should include an analysis of:

1. Occupational illness or injury.
2. First-aid records.
3. Lost time accidents.
4. Workers' compensation claims.
5. Employee medical and hazardous exposure records (Westerkamp, 1995).

Establishing benchmarks will enable an organization to effectively target areas of concern and to find appropriate solutions.

24.4.4 Hazard prevention and control

Lost time and productivity translate into lower production, higher overheads and lower profits, so facility managers should strive for zero tolerance of work-related injuries. Such a programme can save the organization a significant amount of money while increasing employee morale and well-being.

Many tasks put custodians and customers at risk of injury. By identifying potential hazards, developing safety measures and using safety checks, facility managers can minimize the potential for injuries. Among the hazards custodians may be exposed to daily are:

1. *Slips*. This hazard is always present with slippery or wet surfaces and can cause sprained ankles, torn ligaments or knee injuries.
2. *Falls*. Once in an elevated position to change light bulbs, or walking down steps, the risk of falling increases.
3. *Cuts and scratches*. Even the contents of a simple desk-side waste-paper container can expose a custodian to this form of injury.

4. *Back injuries*. Lifting, moving and handling items presents the opportunity for back injuries, ranging from simple muscle strains to herniated discs.
5. *Electrocution*. Such injuries can be caused by frayed wires, presence of water, improperly grounded equipment or equipment that has some electrical malfunction.
6. *Burns*. These can be caused by hot water, steam, hot exhausts, exposed flames or uninsulated components.
7. *Repetitive motion injuries*. These types of injuries result from the consistent motion of parts of the body, such as wrists and hands, particularly when using computer equipment improperly (Bigger and Bigger, 1999).

A company's safety philosophy should put safety first, before productivity, sales or profit. Everyone from the chairman and chief executive officer through the entire organization should believe strongly that accidents, illnesses and injuries are preventable, not inevitable. Like other aspects of the operation, safety is first a management responsibility, but it also requires a close partnership and acceptance of individual responsibility by employees. Management is responsible for instilling the safety philosophy in all employees. Employees have a responsibility to know hazards and use safe practices (Westerkamp, 1996).

Scrutiny of indoor environments continues to increase, and as a result so does pressure on facility managers to address indoor air quality (IAQ) issues proactively. Westerkamp (1998) identifies indoor air pollution as being among the top five environmental public health risks. Indoor air pollutants are two to five times more concentrated – and sometimes more than a hundred times more concentrated – than outdoor pollutants. Building occupants' exposure to these contaminants is higher than ever because many buildings are sealed more tightly, ventilation rates have been cut to save energy, synthetic materials are now incorporated more often in buildings and furnishings, and more personal care products, cleaning chemicals and pesticides are used.

Energy efficiency and indoor air quality traditionally are seen as competing objectives, where one is sacrificed to achieve the other. But through careful planning and creative design work, good indoor air quality doesn't have to come at the expense of energy efficiency (Rospond, 1999). Two main factors should be considered when trying to balance IAQ and energy efficiency: what goes into the building and how the mechanical system is designed to deliver air.

24.4.5 Emergency response

A critical part of any health and safety plan is the procedures to be followed when an emergency occurs. Emergencies may arise from a range of events. Fire is perhaps the most obvious potential emergency that a facility manager should guard against. Facilities contain equipment that can be used in such emergencies, so it is important that occupants know where the equipment is and how it should be used properly. However, other types of emergencies may arise from security breaches or organizational threats, and these command a completely different set of procedures.

Facility managers should not only be aware of issues of continuity planning and disaster recovery, but what to do during an emergency. The main precaution that can be taken is to ensure that adequate procedures exist, that they are disseminated throughout the organization, and that staff are properly informed and trained.

24.4.6 Health and safety training

Since most lost-time injuries occur in the first thirty days of employment (Westerkamp, 1995), new employee safety training is vitally important. This training should include:

1. General safety orientation during hiring interviews.
2. Experienced co-workers going with the new employee on the job and observing safety habits.
3. Job-specific training by giving the worker a job-safety analysis to study before going on the job.
4. Regular follow-up safety training sessions.

To help new employees absorb safety information, the training should be scheduled over several days. Some employers schedule a full week. Retention will be much better if the start-to-finish duration of the instruction is extended. Online self-managed learning packages can also assist.

Health and safety training is a key responsibility of an organization. There is some conflict between responsibilities for this, with both human resource managers and facility managers having a vested interest. This shared interest identifies the synergy between these two essential support services and emphasizes the case for greater integration.

24.5 Regulations

Regulations covering workplace health and safety and proper environmental management are widespread, and of course vary with local conditions. Facility managers must have a library of all health and safety legislation applicable to their facilities and have a good understanding of obligatory responsibilities. Failure to do so will leave the organization open to legal action by affected occupants, which in turn will lead to potentially significant costs and time involvement.

In an increasingly litigious world, facility managers need to ensure that their actions are compliant with statutory requirements. Ignorance is no excuse for negligence. Environmental health and safety is one area that separates the building caretaker, as an employee, from the facility manager, who is now holding a senior position more akin to an employer. While areas of overlap exist between human resource managers and facility managers, there is also the potential for demarcation oversights.

24.6 Preventative action

Prevention is always better than cure. Bernard Williams Associates (1999) identifies a number of important precautions that facility managers should observe in minimizing their exposure to health and safety problems:

1. Resist solutions that lead to workplace overcrowding.
2. Observe appropriate standards of cleaning and maintenance.
3. Use indoor plants to absorb air pollutants.
4. Seek natural lighting and external views from workstations where possible.
5. Avoid the introduction of noise and traffic fumes via open windows in city areas.
6. Choose building and fit-out materials with care, particularly avoiding materials that emit toxic gases.
7. Provide a measure of personal control over environmental conditions.
8. Demonstrate care for the health, safety and comfort of building occupants.
9. Ban smoking within facilities.
10. Do not seal offices unless there is a good, unpolluted supply of natural ventilation.
11. Avoid excessive re-circulation of return air.
12. Keep air humid, but take care in maintaining the humidifying plant free from chemical and bacterial contamination. A suitable level to be maintained for the comfort of workers is between 45% and 55% relative humidity.
13. Watch out for all symptoms of stress.
14. Make sure fire and smoke alarms, any sprinkler systems or other mechanical precautions are regularly tested, and that means of egress are unobstructed.

The list is not exhaustive, but the message is one of minimizing risk by proper care of facilities and compliance with local statutory requirements.

Preventative action has cost rewards. Specific cost reductions from systematic safety planning come from sources that are both direct and indirect. Direct cost impact includes wages paid to injured workers, medical expenses for treatment of any injuries, rehabilitation costs, worker compensation and disability insurance for lost time. Indirect cost impact includes productivity, replacement costs, retraining costs, legal costs arising from injuries, supervisory costs to investigate accidents and incidents, administrative cost of processing claims, in-house medical and rehabilitation costs, wages paid to interrupted workers at the accident scene or afterwards, cost of any emergency capacity or overtime, absenteeism and low morale costs (Westerkamp, 1996).

24.7 Environmental auditing

Environmental auditing is a process of reviewing operating performance and policy with the objective of improving environmental management and associated health and safety. Many organizations are now using environmental auditing as part of their quality assurance procedures. Environmental issues are an emerging area of knowledge for facility managers, and are pervasive to the extent that they affect all stages of the facility life cycle. In the near future facility managers can expect increased standards of environmental compliance and the prospect of significant penalties for breaches and poor performance. Furthermore, environmental efficiency can have positive impact on both profit and worker satisfaction.

The Building Research Establishment (BRE) in the UK has developed an environmental assessment process called BREEAM. This process now comprises three parts. A core assessment of the building fabric and services is carried out in all cases. Two optional parts deal with the quality of the design and its procurement and the management and operating procedures that are used.

The issues assessed as core result in a credible and comparative assessment of the building's potential environmental impacts during operation. This allows buildings of any age to be compared across the range of issues covered to give a consistent tool for the property market. It provides the basis for the determination of an Environmental Performance Index.

The design and procurement option aims to optimize the outcome of a design/procurement exercise. It covers those issues that are of relevance during the design process such as issues of specification and process. The management and operation option provides occupants and managers with an independent audit of the way the building is managed. This presents significant benefits to the organization in financial, legal, health/well-being and image terms. It is intended that this part of the assessment will not only provide a review of performance, but will also lead naturally to the development of an action plan that can be carried forward by the facility manager.

References and bibliography

Alexander, K. (1996). *Facilities Management: Theory and Practice*. E. & F.N. Spon.

Anonymous (1996). 'A Written Safety and Health Plan is an Employer's Best Friend'. *Facilities Design and Management*, 15(10), October, p. 12.

Atkin, B. and Brooks, A. (2000). *Total Facilities Management*. Blackwell Science.

Barrett, P. (1995). *Facilities Management: Towards Best Practice*. Blackwell Science.

Bernard Williams Associates (1999). *Facilities Economics*. Building Economics Bureau Limited.

Bigger, A. S. and Bigger, L. B. (1999). 'Playing It Safe: Emphasising on-the-job safety benefits workers, departments and bottom lines'. *Maintenance Solutions*, July. <http://www.facilitiesnet.com/fn/NS/NS3m9gg.html>

Cotts, D. G. (1999). *The Facility Management Handbook* (2nd edition). AMACOM.

Davies, C. (1996). 'Health and Safety'. *Management Services*, 40(10), October, pp. 22-23.

Green, G. M. and Baker, F. (1991). *Work, Health, and Productivity*. Oxford University Press.

Guzzo, R. A. and Dickson, M. W. (1996). 'Teams in Organizations: Recent research on performance and effectiveness'. *Annual Review of Psychology*, 47, pp. 307-338.

Jones, D. (1997). 'Safety Responsibilities for Facilities Engineers'. *AFE Facilities Engineering Journal*, May/June. <http://www.facilitiesnet.com/fn/NS/NS3a7ea.html>

McGregor, W. and Then, D. (1999). *Facilities Management and the Business of Space*. Arnold.

Moos, R. H. (1996). 'Understanding Environments: The key to improving social processes and program outcomes'. *American Journal of Community Psychology*, 24, pp. 193-201.

Park, A. (1994). *Facilities Management: An Explanation*. Macmillan.

Rondeau, E. P., Brown, R. K. and Lapides, P. D. (1995). *Facility Management*, John Wiley & Sons.

Rospond, K. M. (1999). 'Room for Two: Energy efficiency and IAQ'. *Building Operating Management*, December. <http://www.facilitiesnet.com/fn/NS/NS3b9lb.html>

Salvendy, G. (1997). *Handbook of Human Factors and Ergonomics* (2nd edition). John Wiley & Sons.

Schulze, P. C. (1999). *Measures of Environmental Performance and Ecosystem Condition*. National Academy Press, Washington.

Tompkins, J. A. (1996). *Facilities Planning* (2nd edition). John Wiley & Sons.

Westerkamp, T. A. (1995). 'How Safe is your Department?: Site-specific training provides safeguards against injury – or worse'. *FacilitiesNet*. <http://www.facilitiesnet.com/fn/NS/NS3mi5e.html>

Westerkamp, T. A. (1996). 'Systematic Safety: Guidelines for managing a facility-wide program'. *FacilitiesNet.*
<http://www.facilitiesnet.com/fn/NS/NS3me6h.html>

Westerkamp, T. A. (1998). 'Protecting the Indoor Environment: Surveys can help managers take the lead in preventing problems related to indoor air'. *Maintenance Solutions,* September.
<http://www.facilitiesnet.com/fn/NS/NS3m8id.html>

Zenker, I. (1995). 'Investing in Safety'. *Occupational Hazards*, 57(4), April, pp. 59-61.

Index